SERONO SYMPOSIA PUBLICATIONS FROM RAVEN PRESS
Volume 2

Molecular Biology of Parasites

Serono Symposia Publications from Raven Press
Volume 2

Molecular Biology of Parasites

Editors

John Guardiola, Ph.D.
International Institute of Genetics and Biophysics
Naples, Italy

Lucio Luzzatto, M.D.
International Institute of Genetics and Biophysics
Naples, Italy

William Trager, Ph.D.
The Rockefeller University
New York, New York

Raven Press ■ New York

Raven Press, 1140 Avenue of the Americas, New York, New York 10036

Made in the United States of America

Library of Congress Cataloging in Publication Data
Main entry under title:

Molecular biology of parasites.

(Serono symposia publications from Raven Press; v. 2)
Includes bibliographical references and index.
1. Parasites—Addresses, essays, lectures.
2. Molecular biology—Addresses, essays, lectures.
I. Guardiola, John. II. Luzzatto, L. III. Trager, William, 1910– IV. Series. [DNLM: 1. Parasites—Congresses. 2. Molecular biology—Congresses. W3 SE4779YM / QX 4 M718 1981]
QL757.M55 1983 574.5′249 83-2972
ISBN 0-89004-855-X

Preface

Classically the study of parasites has been concerned primarily with morphological and ecological descriptions. A multitude of unique life cycles have been discovered, which must have played an important role in evolution because of the intimate nature of host–parasite interrelationships. However, these biological systems are generally difficult to experiment with, and the underlying biochemical and physiological phenomena have largely defied in-depth investigation. Molecular biology is only a few decades old, and in that short span of time it has produced spectacular advances in our understanding of the nature of genes, of gene expression, and of the mechanisms of evolution.

The powerful technologies that this new biology has developed, however, have been brought to bear, thus far, only on a limited number of systems. As a crude approximation, one might say that parasitology seeks explanations, while molecular biology is still seeking things to explain.

This admittedly over-simplified view is often misrepresented by regarding molecular biology as "basic" science and parasitology as "applied." This terminology, dangerously suggestive of a hierarchical order, is then tempered by the equally dangerous suggestion that the latter is more relevant than the former. It is true, of course, that there are different ways to approach science. Sometimes a fundamental question is asked—e.g., how does DNA replicate, or how do cells differentiate—and the experimental system is then selected that will best lend itself to pertinent investigation: this is the idea-oriented approach. Others observe natural phenomena and attempt to infer underlying laws from them: this is the descriptive approach. Others yet focus on a particular biological system, one for instance that causes disease, and then seek a way to prevent or cure the disease: this is the problem-oriented approach. There is no doubt that, thus far, parasitology has been primarily descriptive and molecular biology has been more idea-centered: it now seems time to bring the two together with the aim of solving problems.

This volume is a deliberate effort to bring the powerful conceptual and technological tools of contemporary molecular biology to bear on the unique biological systems that parasitology already has indentified and worked out. The volume will be of interest to all parasitologists, immunologists, and molecular biologists.

The Editors

Contents

Contributors

M. Aikawa
Institute of Pathology
Case Western Reserve University
Cleveland, Ohio 44106

P. A. Battaglia
Istituto Superiore di Sanita
00161 Rome, Italy

P. Bazzicalupo
International Institute of Genetics and Biophysics
80125 Naples, Italy

A. Bernards
Section for Medical Enzymology and Molecular Biology
Laboratory of Biochemistry
University of Amsterdam
Jan Swammerdam Institute
1005 GA Amsterdam
The Netherlands

N. Bone
Department of Molecular Biology
University of Edinburgh
Edinburgh EH9, Scotland

P. Borst
Section for Medical Enzymology and Molecular Biology
Laboratory of Biochemistry
University of Amsterdam
Jan Swammerdam Institute
1005 GA Amsterdam
The Netherlands

A. Cascino
International Institute of Genetics and Biophysics
80125 Naples, Italy

D. Cioli
Laboratory of Cell Biology
National Research Council
00196 Rome, Italy

M. del Bue
Istituto Superiore di Sanita
00161 Rome, Italy

E. Dore
Laboratory of Cellular Biology and Immunology
Istituto Superiore di Sanita
00161 Rome, Italy

M. J. Friedman
Cancer Research Institute
University of California
San Francisco, California 94143

C. Frontali
Laboratory of Cellular Biology and Immunology
Istituto Superiore di Sanita
00161 Rome, Italy

T. Gibson
Department of Molecular Biology
University of Edinburgh
Edinburgh EH9, Scotland

M. Goman
Department of Molecular Biology
University of Edinburgh
Edinburgh EH9, Scotland

J. Guardiola
International Institute of Genetics and Biophysics
80125 Naples, Italy

A. M. Guerrini
International Institute of Genetics and Biophysics
80125 Naples, Italy

J. E. Hyde
Department of Molecular Biology
University of Edinburgh
Edinburgh EH9, Scotland

M. Iaccarino
International Institute of Genetics and Biophysics
80125 Naples, Italy

G. W. Langsley
Department of Molecular Biology
University of Edinburgh
Edinburgh EH9, Scotland

L. Luzzatto
International Institute of Genetics and Biophysics
80125 Naples, Italy

J. S. McBride
Department of Zoology
University of Edinburgh
Edinburgh EH9, Scotland

M. P. Nuti
Institute of Microbiology
University of Pisa
Pisa, Italy

M. Ottaviano
Istituto Superiore di Sanita
00161 Rome, Italy

M. Ponzi
Istituto Superiore di Sanita
00161 Rome, Italy

J. G. Scaife
Department of Molecular Biology
University of Edinburgh
Edinburgh EH9, Scotland

J. Schell
Laboratorium voor Genetika
Rijksuniversiteit Ghent
Belgium

W. Trager
The Rockefeller University
New York, New York 10021

D. Walliker
Department of Genetics
University of Edinburgh
Edinburgh EH9, Scotland

L. Willmitzer
Max-Planck-Institut für Züchtungsforschung
Köln-Vogelsang
Federal Republic of Germany

R. K. S. Wood
Imperial College
London SW7 2BB
United Kingdom

N. K. Yankofsky
Department of Molecular Biology
University of Edinburgh
Edinburgh EH9, Scotland

J. W. Zolg
Department of Molecular Biology
University of Edinburgh
Edinburgh EH9, Scotland

Introduction

William Trager

Rockefeller University, New York, New York

Parasitism involves an intimate association between two different kinds of organisms—one, the host, providing food and shelter for the other, the parasite. The host may or may not be injured by the parasite. It may soon expel the parasite, or it may harbor it for many years. Since the parasite cannot exist in nature without its host, it is not to the parasite's advantage to destroy its host. At least it must not destroy it until ready to move to another. Some hosts are benefitted by certain parasites, and some actually depend on their parasites in a special type of association called mutualism. Such mutualistic or symbiotic associations may have been at the origin of chloroplasts and mitochondria, and so at the basis of most eukaryotic cells.

Throughout the living world, from prokaryotes to man, parasitic associations are very common. There is no organism (except for viruses) that does not have its parasites. Furthermore, parasitic organisms are found in all major taxonomic groups. To study parasites as organisms in their own right is relatively simple and straightforward, but to study the interrelations between the parasite and its host, i.e., to study parasitism, requires all the disciplines of biology from ecology to biophysics. It is this approach, studying the physiology and biochemistry and cell biology of host–parasite relationships, that will constitute the main body of future work in parasitology, and it is this approach that has been taken in this volume.

A complete study of the molecular biology of parasites would begin with a discussion of the establishment of infection—what factors enable the parasite to recognize and enter an appropriate host, and, equally important, what factors in the host permit it to accept the parasite. Some parasites enter hosts as resistant or dormant forms, for example, the cysts of intestinal amoebae or the eggs of *Ascaris*. Here the physiochemical factors of interest are those that trigger excystation or hatching. Others may be ingested into a vector host as active forms sucked up with the blood, as when trypanosomes are taken up by a tsetse fly or malaria parasites by a mosquito. The new environment induces special cycles of differentiation culminating in forms again able to initiate new infections when they are injected into a new definitive host by the biting insect.

Once within the host, a variety of organ and tissue tropisms come into play. Thus, certain larval trematode worms find their way to the eyes of a fish no matter where they enter its body. Intracellular parasites must have special recognition sites to permit engulfment and uptake of the parasite by an appropriate cell, as must their corresponding host cells. The most beautiful examples here are provided by the different species of malarial parasites and the different kinds of erythrocytes they can enter, many showing exquisite specificity.

In these early interactions, the parasite must have ways of coping with the strong ability of all organisms to reject foreign structures, living or dead. I find it fascinating that a large group of parasitic protozoa, the leishmanias, has subverted this ability; these organisms actually parasitize macrophages, cells that are normally in the first line of defense against invaders. A variety of biochemical and immunological factors may affect innate resistance in these early interactions, and all of these are determined by the genetic makeup of the parasite as well as the host.

Once the parasite is in its host, we would want to consider what occurs at the parasite–host interface, the roles of surfaces and membranes in sheltering and nourishing the parasite. We are here very much at the beginning of what should be a long and exciting future story. With intracellular parasites, what is the significance of the parasitophorous vacuole within which some of them lie? Why is a malarial parasite within an erythrocyte surrounded not only by its plasma membrane but also by the parasitophorous membrane derived originally from the host cell, whereas a babesia lies directly in the red cell cytoplasm, separated only by its plasma membrane? What differences in nutritional physiology and in metabolism result from these architectural differences? Such discussions lead into the whole subject of nutritional and metabolic and hormonal interrelations between host and parasite including consideration of effects of host nutrition on the parasite, the modes of uptake of nutrients by the parasites, and their nutritional requirements. *In vitro* cultivation of the parasite is of special importance here, in particular axenic cultures in defined media. The energy metabolism of parasites is often somewhat different from that of their hosts, providing opportunities for interference by chemotherapy. Such is also the case with the biosynthetic capabilities of the parasites.

Modification of a host by an established parasite, and the host's reaction to it, is a major aspect of great practical importance, since it involves both the nature of parasitic disease and of acquired immunity to parasites. Mutualism is also a special aspect of this large subject. Parasites affect their hosts in various ways, both nonspecific and specific. The sparganum of the cestode *Spirometra* produces a growth factor that causes host rats to grow to gigantic size. Intracellular parasites, bacterial as well as protozoan, often induce marked changes in the host cell. Thus, with malaria parasites the permeability of the erythrocyte plasma membrane is increased and parasite antigens appear in the host cell plasma membrane. Even more striking is the extensive hypertrophy of the host cell induced by some intracellular parasites, usually accompanied by polyploidy. Such an effect reaches an extreme with certain microsporidian parasites of fish, where there arise great tumorous formations, or xenomas.

Infectious disease, or the mechanisms of pathogenesis by infectious agents, remains an elusive subject. There may be direct tissue destruction or injurious effects by toxins but usually the pathogenic effects are more subtle and represent the combination of a variety of different factors, including immunological ones. One of the most interesting aspects of the study of acquired immunity to parasites is how parasites evade the hosts' immune response. One popular way is for the parasite to acquire a coating with host serum proteins. In this way, adult schistosomes can thrive for years in a host immune to superinfection by newly invading larval schistosomes. Another method of evasion is antigenic variation, beautifully illustrated in the surface coat glycoprotein of the African trypanosomes. The protein portion of this glycoprotein can vary considerably. Hence by the time a host has developed adequate antibodies to the coat proteins of an initial population of trypanosomes, a variant population appears with a different coat protein. Subsequently, when the host has developed antibodies, a third population with still a third variant coat grows out, and so on. The intracellular location is still another way for a parasite to be protected from host antibodies. *Plasmodium falciparum* of man and some other malarial parasites of monkeys have developed a way to avoid going through the spleen. By the time the parasite is half grown and therefore likely to be detected and removed by the spleen, small protrusions, which are evidently sticky, appear on the host cell membrane and attach the infected cell to the endothelial capillary walls of the brain, heart, and other organs. Hence these larger parasites, during the latter half of their 48-hour cycle, do not circulate through the spleen. It would be of great interest to understand fully the molecular and cell biology of this sticking process. Immunosuppressive effects are common with different kinds of infectious agents. Despite all these mechanisms for evasion, hosts do acquire immunity. Immunoparasitology with a special view to the development of effective vaccines is a growing field.

Some of the parasitic associations that are mutualistic and do not involve injury to the host are the nitrogen-fixing bacteria in the roots of plants, the bacteroids of roaches, tsetse flies and lice, the cellulose-digesting protozoa of termites, the intracellular algae of hydra and many other marine invertebrates.

Many of the eukaryotic parasitic organisms—the fungi, protozoa, and helminths—have complex life cycles involving two and sometimes three different kinds of hosts. Here, the parasites must not only be adapted for development in each kind of host, they must also produce at appropriate times special stages able to effect the transfer from one host to another, or from one environment to another.

The African trypanosomes again provide a good example. These flagellates live in the blood or tissue spaces of man and other mammals and have a distinctive metabolism which does not involve a cytochrome-mediated respiratory chain. Their mitochondrial structure is very primitive and lacks cristae. However, some stumpy-looking forms that have the beginnings of mitochondrial differentiation, and which are specially adjusted for development in the midgut of a tsetse fly appear in the blood. There, they transform into procyclic trypanosomes characterized by a cytochrome-mediated respiration and (and this is most interesting) by complete loss of infectivity for a vertebrate host. After several weeks, trypanosomes invade the salivary glands or proboscis of the fly, where a third cycle of multiplication and development ensues that culminates in forms very like the bloodstream forms. These forms initiate infection if injected by the fly into a vertebrate host. What the environmental factors are that trigger these cycles and how they activate particular genes to be expressed are subjects now under study. Clearly they are of basic biological significance and require the application of the tools of molecular biology.

Molecular Biology of Parasites, edited by
J. Guardiola, L. Luzzatto, and W. Trager.
Raven Press, New York © 1983.

Host–Parasite Interaction: Electron-Microscopic Study

Masamichi Aikawa

Institute of Pathology, Case Western Reserve University, Cleveland, Ohio 44106

Host–parasite interaction is important to the parasite protozoa, since their survival depends on the host cells that supply the environmental and nutritional requirements. They cannot live away from their host cells or host cell nutrients. Intracellular protozoa are equipped with several specialized features that enable them to utilize other living cells as an environment.

There are many gaps in our knowledge of the specific mechanisms that are involved in the interaction between parasites and their hosts. The early morpho-

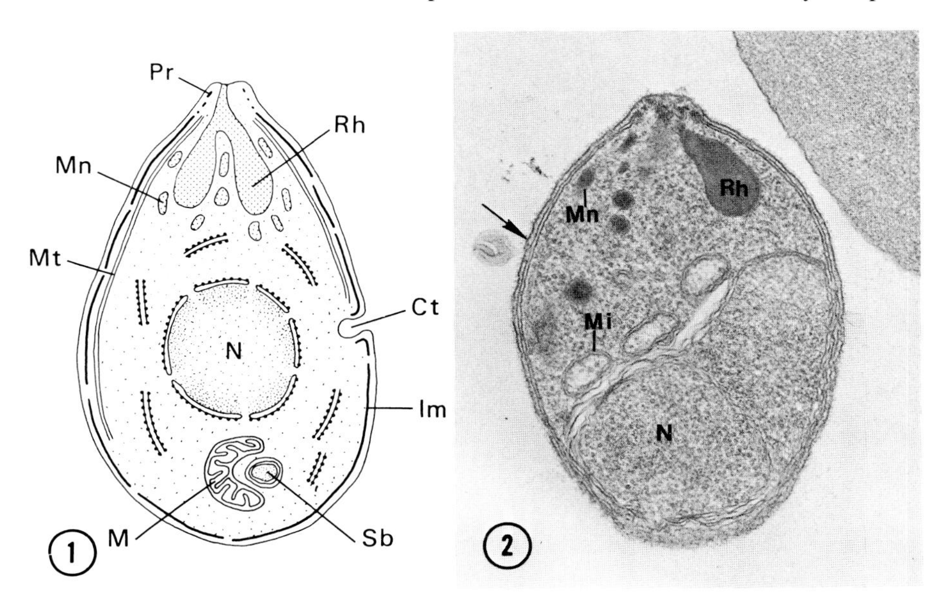

FIG. 1. Schematic diagram of the merozoite of *Plasmodium.* Ct, cytostome; Im, inner membrane; M, mitochondria; Mn, micronemes; Mt, microtubule; N, nucleus; Pr, polar rings; Rh, rhoptry; and Sb, spherical body.
FIG. 2. Free merozoite of *Plasmodium* showing rhoptry (Rh), micronemes (Mn), a nucleus (N), mitochondria (Mi), and a surface coat *(arrow).* $\times$43,000. (From Aikawa and Kilejian, ref. 1, with permission.)

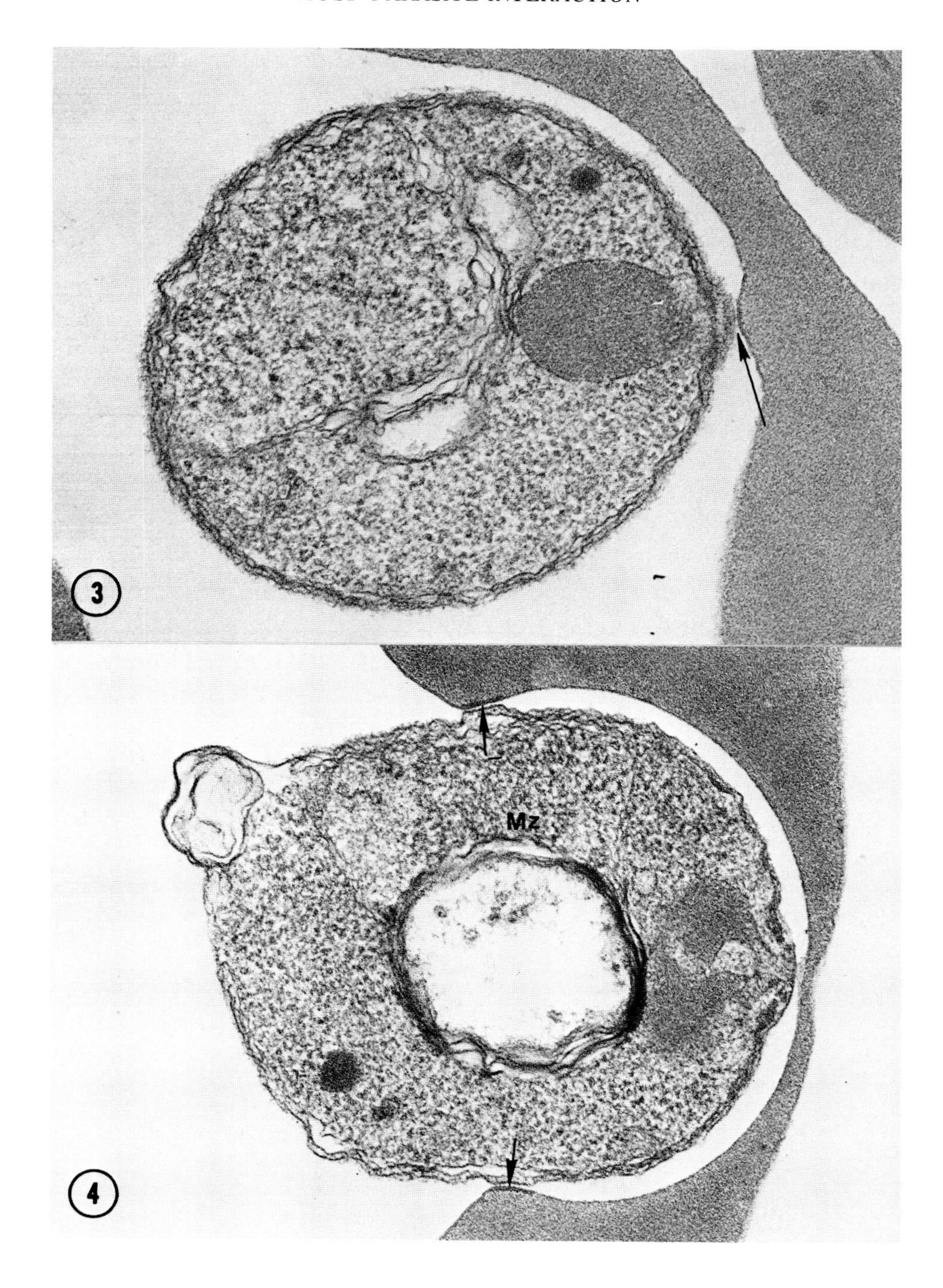
3
4
Mz

logical study on host–parasite interaction was affected by the limited resolving power of the optical light microscope and failed to reveal the detailed structural interaction between the hosts and parasites. The recent advent of electron microscopy, including thin-sectioning, freeze-fracture, and scanning electron microscopy, has contributed greatly to the understanding of host–parasite interaction. However, available information is still limited to the few protozoa that can be studied under *in vitro* conditions, since technical difficulties prevent detailed observations of such interactions *in vivo*. These parasites include members of the sporozoa such as *Nosema*, *Plasmodium*, *Eimeria*, *Toxoplasma*, and *Babesia* and two flagellates, *Leishmania* and *Trypanosoma* (1). There are three major features involved in the interaction of these protozoa and their host cells: (a) entry into the host cells, (b) growth of intracellular protozoa within the host cells, and (c) alteration of the host cells. These three topics will be discussed in this chapter.

HOST CELL ENTRY BY PROTOZOA

Generally speaking, there are three possible pathways that an intracellular parasite can use to gain entry into the host cell: (a) direct passage through the host cell membrane, (b) fusion of the host and parasite membranes, and (c) endocytosis. The information that has accumulated in recent years now demonstrates that with the exception of *Nosema*, all of the protozoa studied so far appear to enter cells by endocytosis. *Nosema*, a member of the microsporidia, is uniquely specialized in its capacity to penetrate the host plasma membrane (28). Its spores possess a complex extension apparatus that discharges a tube with explosive force. The discharged tube penetrates the host cell membrane and serves as a vehicle for the transfer of the sporoplasm from the spore to the host.

No generalization of specific mechanisms of endocytotic entry of parasitic protozoa can be made because there are many variables among protozoa. However, endocytosis generally involves three processes, including (a) initial attachment of the protozoa to the host cell membrane, (b) invagination of the host cell membrane, and (c) sealing of the host cell membrane after completion of the protozoa entry. In this section, these processes will be discussed by demonstrating host cell entry by *Plasmodium* and *Toxoplasma* electron microscopically.

Interaction Between Malarial Merozoites and Erythrocytes

Since the understanding of the structure of *Plasmodium* is essential in analyzing the interaction between the erythrocytes and merozoites, a brief description of the merozoite is illustrated in Figs. 1 and 2. The parasite is elongated in shape and is

FIG. 3. A merozoite of *P. knowlesi* at the initial contact *(arrow)* between the merozoite's apical end and an erythrocyte. ×45,000. (From Aikawa et al., ref. 6, with permission.)

FIG. 4. Erythrocyte entry by a merozoite (Mz). The junction *(arrow)* is formed between the thickened membrane of the erythrocyte and a merozoite membrane. ×40,000. (From Aikawa et al., ref. 6, with permission.)

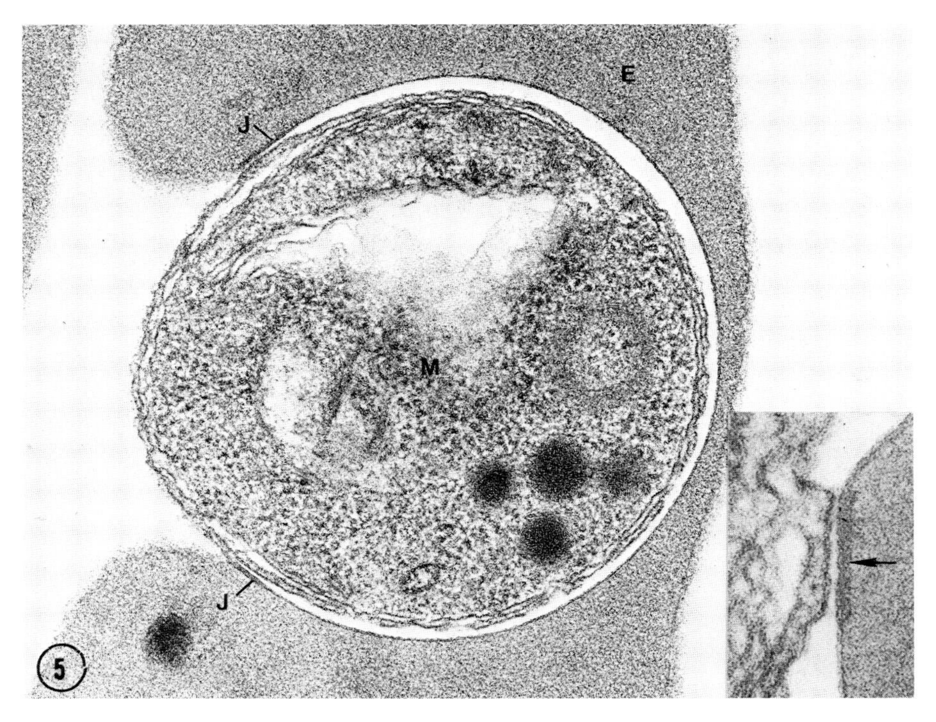

FIG. 5. Advanced stage of erythrocyte (E) entry by a merozoite (M). The junction (J) is always located at the orifice of the erythrocyte entry. ×43,000. **Inset:** High magnification of the junction showing thickening of the erythrocyte membrane where the merozoite is attached *(arrow)*. ×128,000. (From Aikawa et al., ref. 7, with permission.)

surrounded by a pellicle complex of two membranes and a row of subpellicular microtubules. The lateral side of the pellicle shows a circular indentation called the cytostome. This structure will be engaged in the ingestion process of host cell cytoplasm in the later stages. The apical end is a truncated cone-shaped projection demarcated by polar rings. Electron-dense rhoptries and micronemes are present in the apical end of the parasite, and their ductules extend to the tip of the apical end. The nucleus is usually located in the midportion. Anterior to the nucleus is a cluster of vesicles and membranous lamellae that appear to be the Golgi complex. The posterior portion is occupied by a mitochondrion and a few electron-dense inclusions.

Even though specific reciprocal recognition sites on protozoa or their host cell surfaces have not been characterized, indirect experimental evidence suggests the involvement of such ligands for *Plasmodium*. Alteration of the host plasma membrane by proteases has been shown to abolish specifically *Plasmodium* attachment to the erythrocytes (19). Trypsinization of human erythrocytes markedly reduces their infectivity for *P. falciparum* but does not alter the penetration by *P. knowlesi*.

FIG. 6. Freeze-fracture electron micrograph showing that the merozoite (M) is creating a slight invagination of the erythrocyte membrane during the initial stage of invasion. A small indentation *(arrow)* is present at the apical end. ×50,000. (From Aikawa et al., ref. 7, with permission.)

In contrast, chymotrypsin treatment of the erythrocytes blocks the infection by *P. knowlesi* but has no influence on the infection by *P. falciparum*.

Similar experiments have been performed for *Trypanosoma* and *Babesia*. Treatment of macrophages with trypsin and chymotrypsin does not influence their ability to ingest sensitized sheep erythrocytes but abolishes the attachment of trypomastigotes and epimastigotes of *Trypanosoma cruzi* to these host cells. This study by Nogueria and Cohn (22) indicates that macrophage receptors of C_3 complement components are not involved in the attachment of *T. cruzi*, since chymotrypsin does not alter these receptors. Unlike *T. cruzi*, C_3 and C_5, as well as factors of the alternative complement pathway, are required for the penetration of *Babesia* into the erythrocytes (10). However, it is unclear whether complement components participate in the attachment of *Babesia* to the erythrocytes or in subsequent steps in endocytosis.

More evidence of the presence of host recognition sites comes from the observation that for successful endocytosis, the merozoite must come in contact with the

FIG. 7. Freeze-fracture electron micrograph showing the P face (Pv) of a parasitophorous vacuole. The P face (Pe) of the erythrocyte membrane at the neck of the invagination is covered with IMP, but they disappear beyond the point where the neck of the invagination abruptly expands. ×41,000. **Inset:** High magnification of the P face (Pe) of the erythrocyte membrane at the neck of the invagination. A band of arrayed particles *(arrow)* is seen at the point just before the neck of the erythrocyte membrane invagination expands into the parasitophorous vacuole. ×68,000. (From Aikawa et al., ref. 7, with permission.)

erythrocytes with its apical end, which is characterized by the presence of specialized organelles (6). However, it is not known what makes the exposed elements on the apical end different from the rest of the merozoite surface. A possible factor underlying such attachment may be the difference in surface charge of the host and merozoite. Cytochemical studies, using positively charged colloidal ions, have indicated that unlike the host erythrocyte membrane, the merozoite's membrane lacks exposed sialic acid groups (24). In addition, the parasite extracts contain about half of the amount of sialic acid per unit weight when compared to the red cell extracts. Recently, Miller et al. (19) reported that the initial recognition and attachment between the merozoites and erythrocytes involves specific determinants associated with Duffy-blood-group-related antigen.

When the apical end of the merozoite contacts the erythrocyte membrane, it becomes slightly raised initially at the interaction point (Fig. 3), but eventually a depression is created in the erythrocyte membrane (6). The membrane to which the parasite is attached becomes thickened, measuring about 150 Å in thickness, whereas the normal erythrocyte membrane measures 75 Å in thickness. This thickened erythrocyte membrane forms a junction with the parasite membrane. As the invasion progresses, the depression in the erythrocyte deepens and conforms to the shape of

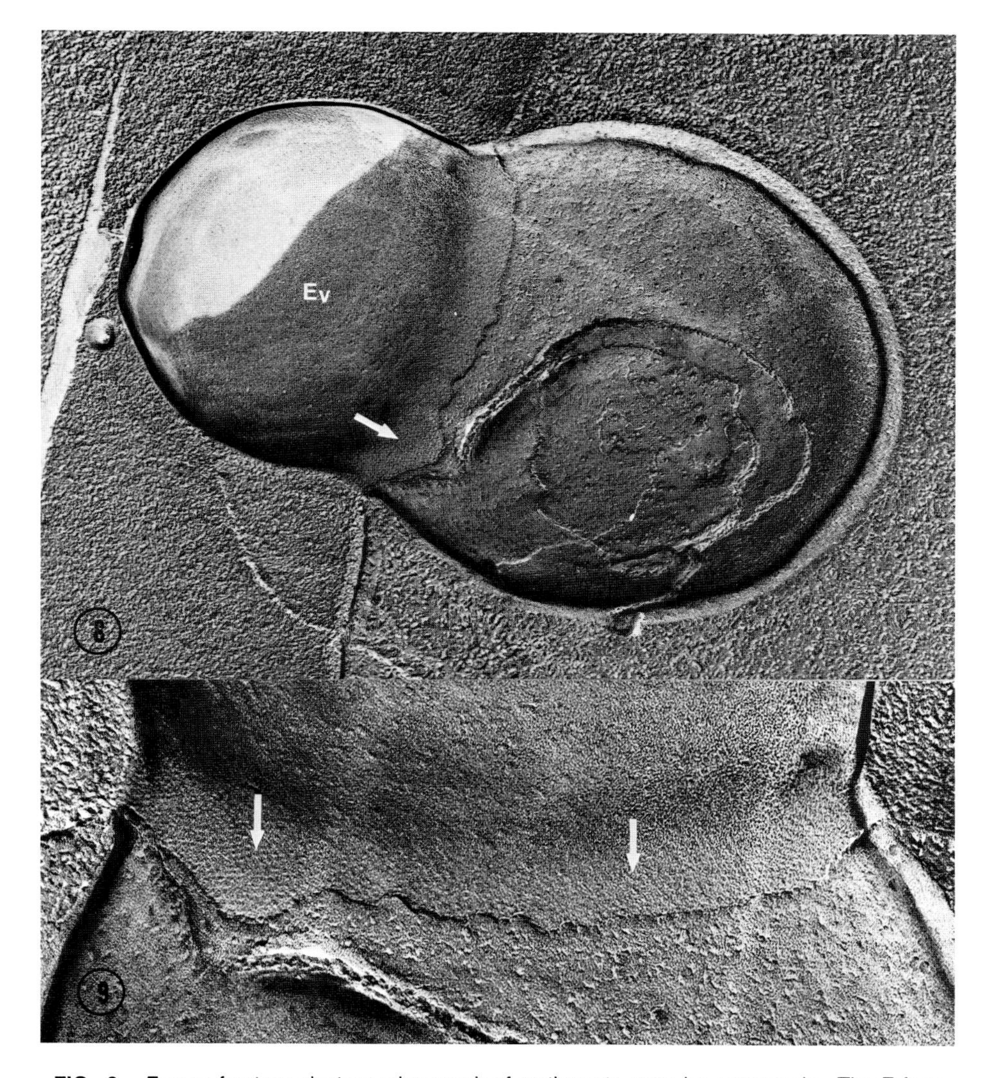

FIG. 8. Freeze-fracture electron micrograph of erythrocyte entry by a merozoite. The E face of the erythrocyte membrane at the neck of the invagination consists of a narrow circumferential band of arrayed pits *(arrow)*. (Ev) is the E face of the vacuole membrane. ×45,000. (From Aikawa et al., ref. 7, with permission.)
FIG. 9. High-magnification micrograph showing a band of arrayed pits *(arrow)* at the neck of the erythrocyte invagination. ×64,000. (From Aikawa et al., ref. 7, with permission.)

the merozoite (Fig. 4). At this time, the thickened, electron-dense zone on the erythrocyte membrane, forming a junction with the merozoite, is no longer observed at the initial attachment point, but now appears at the orifice of the merozoite-induced invagination of the erythrocyte membrane (6) (Fig. 5).

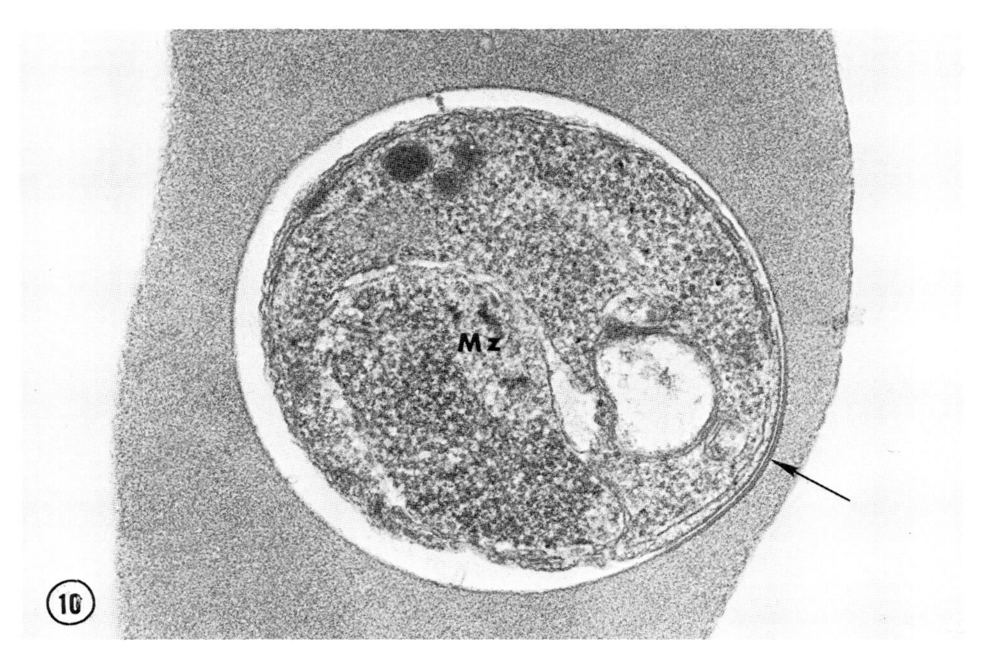

FIG. 10. Merozoite (Mz) inside of an erythrocyte. However, the posterior end of the merozoite is still attached *(arrow)* to the thickened erythrocyte membrane. ×40,000. (From Aikawa et al., ref. 6, with permission.)

The thickened area of the erythrocyte membrane measures 15 μm in thickness and 250 μm in length, which appears to be due to the thickening of the erythrocyte bilayer's inner leaflet (Fig. 5). The junctional gap between the erythrocyte membrane and the merozoite membrane is about 10 μm, and the fine fibrils extend between these two parallel membranes. Since the junction is always located at each side of the orifice regardless of the plane through which the section is cut, the junction appears to be circumferential in shape. However, since thin-section electron microscopy cannot reveal the details of the junction structure, freeze-fracture is the only method for the analysis of the membrane structure.

Freeze-fracture of the erythrocyte membrane shows many intramembrane particles (abbreviated as IMPs) on the P face and a few IMPs on the E face (7). As the merozoite invasion progresses, the invagination of the erythrocyte membrane is observed (Fig. 6). At the point just before the neck of the invagination abruptly expands into the parasitophorous vacuole, a narrow band of rhomboidally arrayed particles is seen on the P face of the erythrocyte membrane (7) (Fig. 7). Matching rhomboidally arrayed pits are seen on the E face (Figs. 8 and 9). This circumferential band corresponds to the junction region between the erythrocyte and merozoite membranes observed by thin-section and disappears at the point of expansion of the vacuole. The P face of the erythrocyte at the neck of the invagination is covered

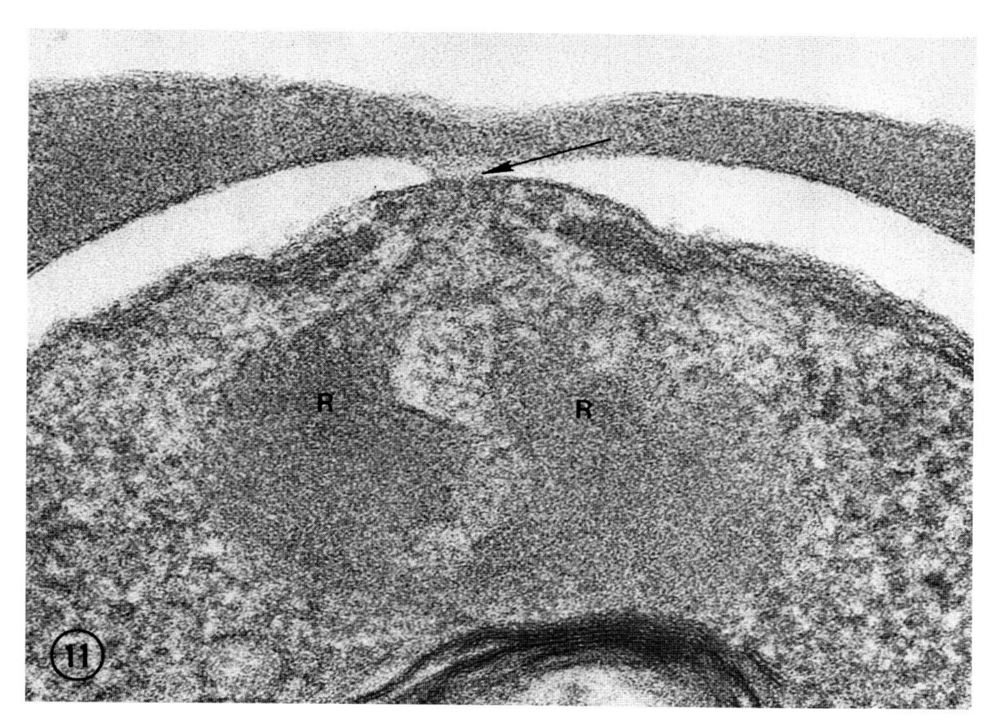

FIG. 11. High-magnification electron micrograph showing two rhoptries (R) at the apical end. An electron-opaque projection connects the apical end and the erythrocyte membrane *(arrow)*. ×108,000. (From Aikawa et al., ref. 6, with permission.)

with IMPs as the normal erythrocyte membrane. However, IMPs abruptly disappear beyond the junction (Fig. 7).

About 30% of the erythrocyte membrane protein consists of spectrin and actin, which form a lattice-like contractile system located on the inner aspect of the erythrocyte membrane (21,26). The arrangement of the contractile protein into a network is thought to facilitate changes in the shape of the erythrocyte (25). Therefore, it is possible that the presence of the dense zone at the junction site may represent contractile protein aggregates (7). Rhomboidally arrayed pits on the E face seen by freeze-fracture may be the points where the contractile proteins anchor to the transmembrane proteins in the erythrocyte membrane.

The circumferential band disappears at the parasitophorous vacuole expansion point. There is a significant difference noted between the number of IMPs present on the P faces of the erythrocyte and vacuole membranes (7). On the other hand, no obvious difference was noted between the IMPs on the E faces of the erythrocyte and vacuole membranes.

The contents of rhoptries, which are located in the parasite's apical region, flow and diffuse into the erythrocyte membrane and may be involved in the formation

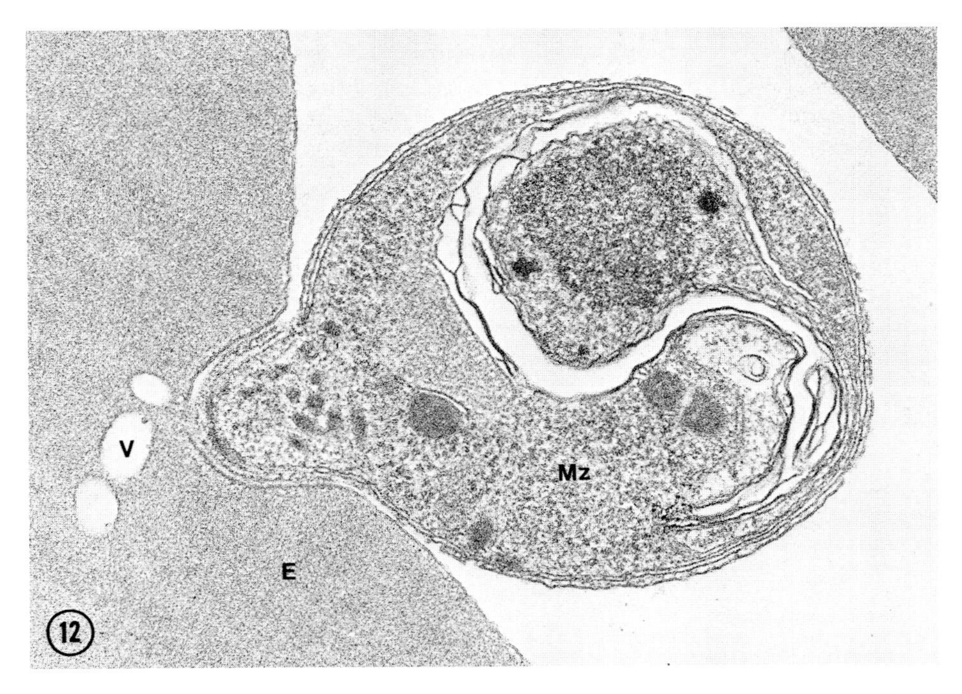

FIG. 12. An attachment between the apical end of a cytochalasin B-treated merozoite (Mz) and rhesus erythrocyte (E). A few vacuoles (V) are seen in the erythrocyte cytoplasm near the attachment site. ×64,000. (From Aikawa et al., ref. 7, with permission.)

of a relatively protein-free vacuole membrane. Possible mechanisms might include the crosslinking of spectrin, phospholipase activity, or actual creation of phospholipid bilayers by the rhoptries (7).

When the erythrocyte entry by a plasmodial merozoite is almost completed, a small orifice is seen at the posterior end of the erythrocyte and the junction moves closer to the merozoite's posterior end. When the entry is completed, the junction appears to fuse at the end (Fig. 10), closing the orifice in the fashion of an iris diaphragm (6). The merozoite membrane still remains in close apposition to the thickened erythrocyte membrane at the final closure point.

The events that occur during invasion relate to the endocytosis process by which phagocytic cells ingest particles, other cells, and microorganisms. Griffin and his associates (13) proposed the hypotheses for endocytosis of particles, namely, specific attachment triggering endocytosis and zippering. Triggering requires specific receptors for attachment, but ingestion is independent of receptors outside of the attachment. Zippering is attachment to the receptors around the circumference of the particles and requires a metabolically active cell. However, neither model appears to explain the observation during invasion by *Plasmodium*. For the erythrocyte invasion by the malarial merozoites, movement of the junction appears to occur at the membrane level; this movement may be related to the lateral displace-

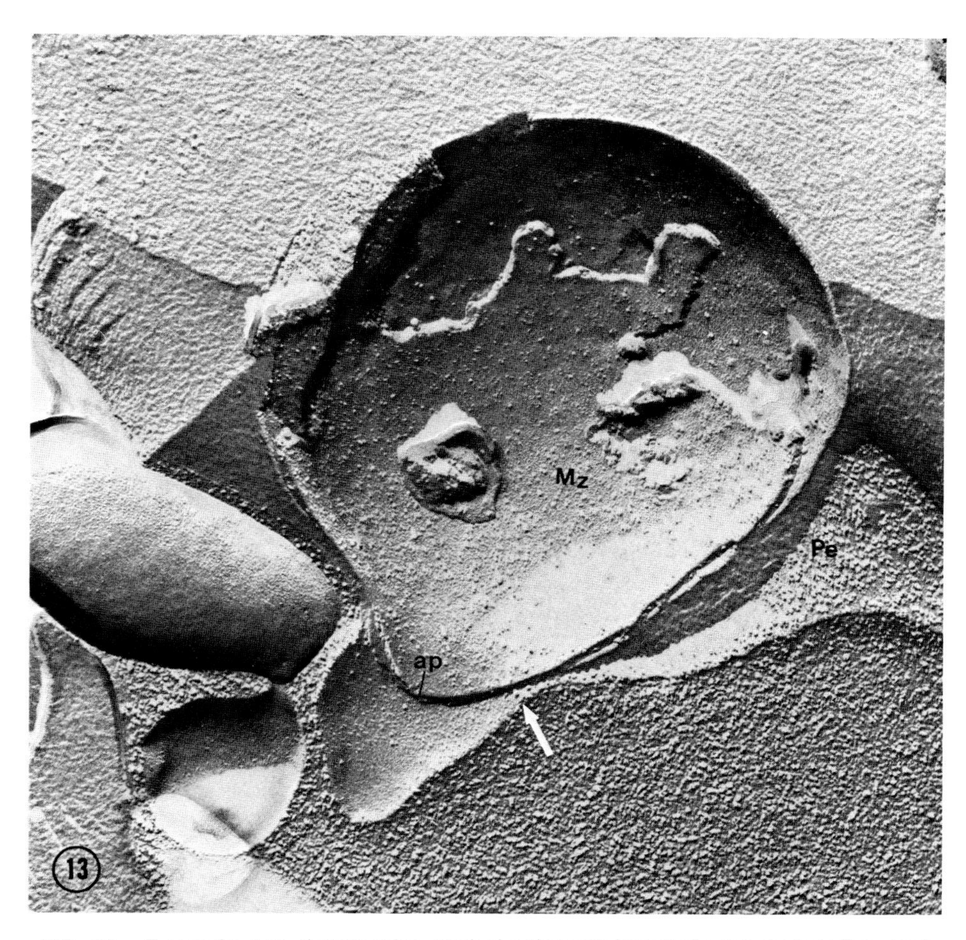

FIG. 13. Freeze-fracture electron micrograph showing attachment of a cytochalasin B-treated merozoite (Mz) to the erythrocyte membrane. The invagination of the erythrocyte membrane extends beyond the apical end (Ap) of the merozoite. IMPs on the P face (Pe) of the erythrocyte membrane at the end of the invagination neck *(white arrow)* abruptly disappear. × 50,000. (From Aikawa et al., ref. 7, with permission.)

ment of the junction by the membrane flow agency. Similar junctions have also been found during erythrocyte entry by *Babesia*, indicating that *Babesia* probably enters host cells by a mechanism similar to that demonstrated for *Plasmodium* (1).

In the early 1960s, investigators using electron microscopy reported that plasmodial merozoites possess rhoptries and micronemes that were associated with host cell entry. Throughout invasion, the apical end remains in contact with the erythrocyte membrane (Fig. 11). There is an electron-dense band between the tip of the apical end and the erythrocyte. The band appears to be continuous with a common duct of the rhoptries. During invasion, the lower electron density in the duct suggests a release of the rhoptry contents. Kilejian (15) suggested that the rhoptries and

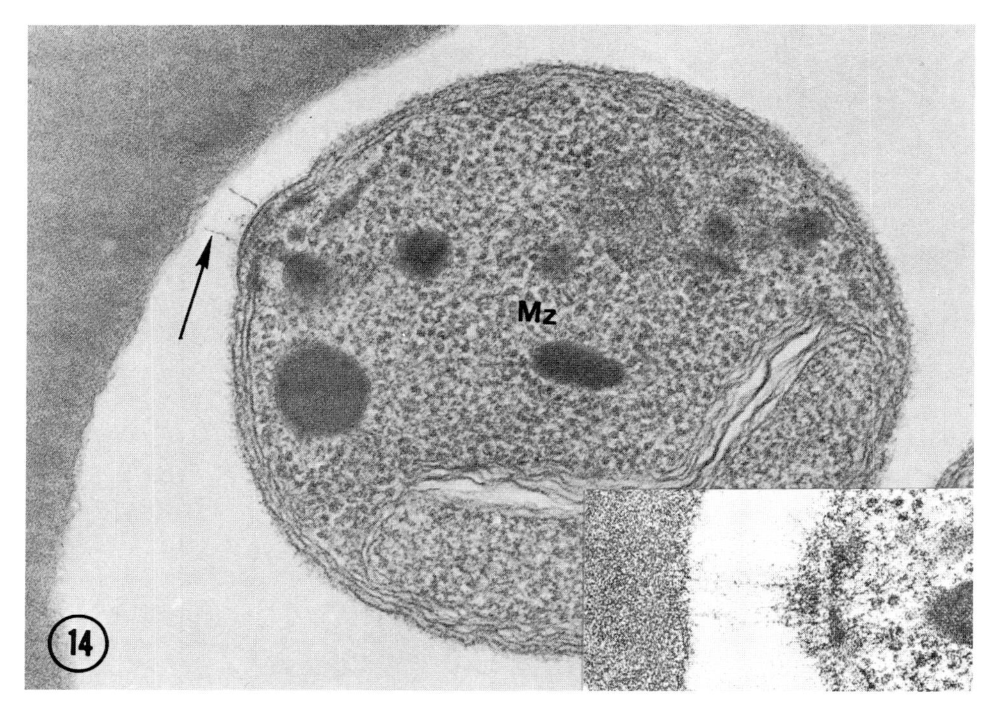

FIG. 14. A cytochalasin B-treated merozoite (Mz) is connected with a Duffy-negative human erythrocyte by two fine fibrils *(arrow)* that are extending from the edge of the apical end. ×50,000. **Inset:** Two fibrils connecting the apical end of a merozoite and a Duffy-negative erythrocyte. ×78,000. (From Miller et al., ref. 20, with permission.)

micronemes of an avian malarial parasite *P. lophurae* contain a histidine-rich protein that assists to invaginate the erythrocyte membrane. Our observation on the connection between the rhoptries and the erythrocyte membrane presented here supports the supposition that the rhoptries play a role in merozoite entry into the erythrocyte. However, the identification of the function of rhoptries and micronemes must await the isolation of contents of these organelles for analysis of their chemical and physical properties.

Interaction Between Cytochalasin-Treated *Plasmodia* and Erythrocytes

The isolation of the attachment phase of protozoa to host cells can be achieved by using cytochalasin B. Cytochalasin B is known to affect microfilaments and glucose transport of cells. Cytochalasin B-treated plasmodial merozoites attach to the erythrocyte but do not invade the erythrocyte (Fig. 12).

When cytochalasin B-treated merozoites of *P. knowlesi* are incubated with rhesus erythrocytes, Duffy-positive human erythrocytes, and Duffy-negative human erythrocytes, light microscopy shows that the apical ends of the merozoites are oriented

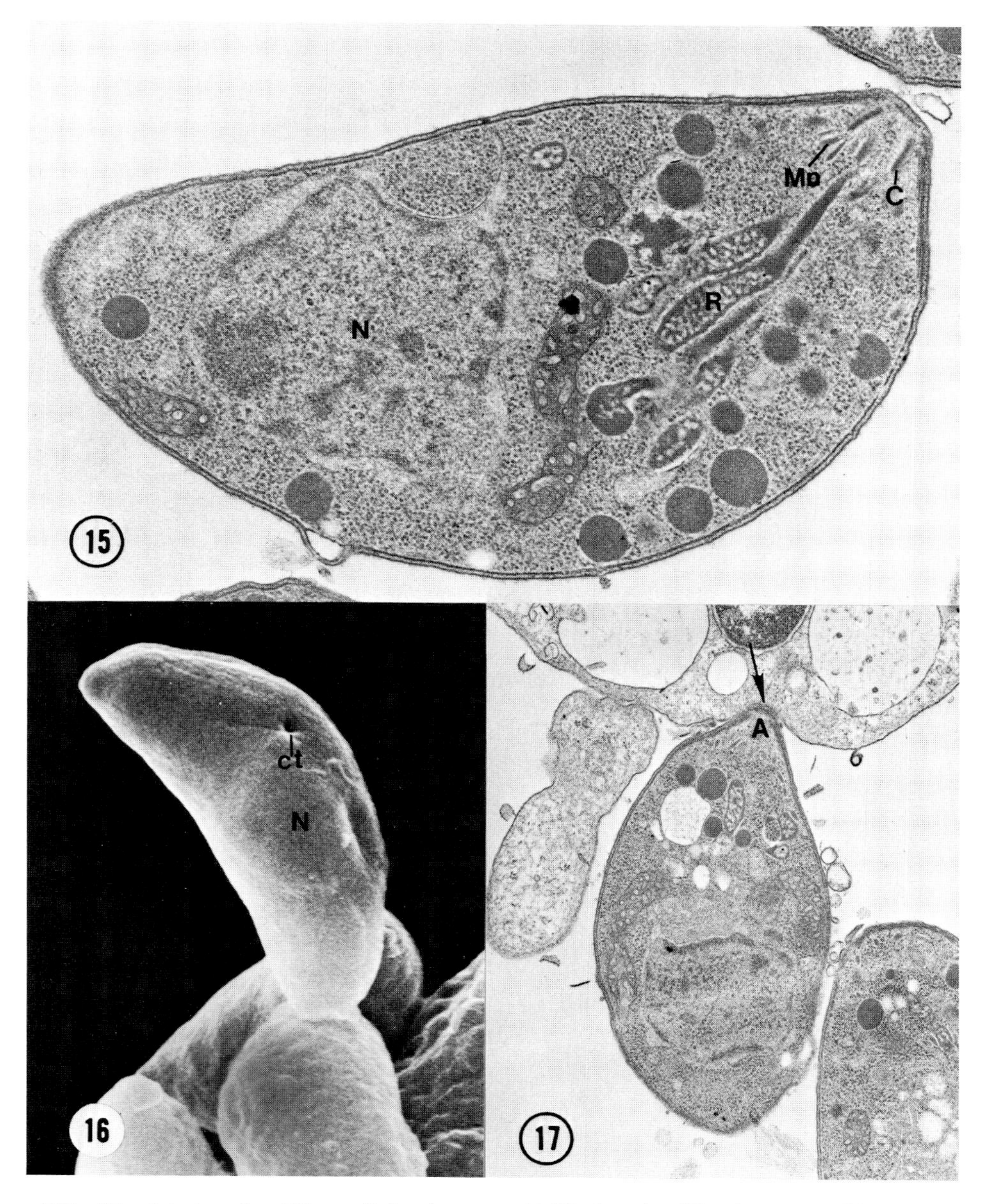

FIG. 15. A tachyzoite of *T. gondii* showing a conoid (C), rhoptries (R), micronemes (Mn), and a nucleus (N). ×27,000. (From Aikawa et al., ref. 5, with permission.)
FIG. 16. Scanning electron micrograph of *T. gondii* showing a cytostome (Ct) and a nuclear region (N). ×22,000. (From Aikawa et al., ref. 5, with permission.)
FIG. 17. Electron micrograph of *T. gondii* with its anterior end (A) directed toward a host cell. A small invagination *(arrow)* is formed at the surface of the host cell. ×14,000. (From Aikawa et al., ref. 5, with permission.)

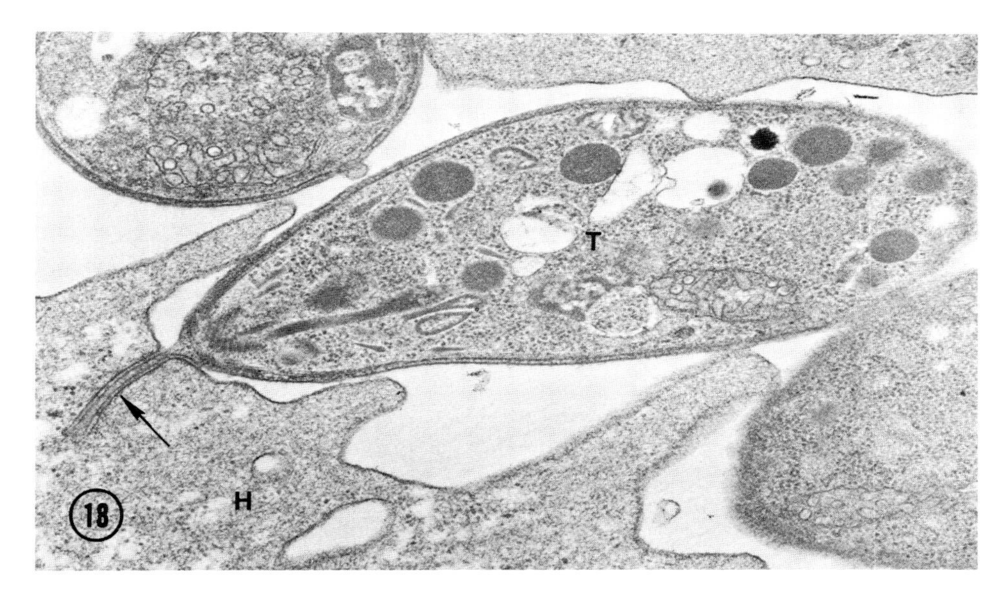

FIG. 18. A tachyzoite (T) showing a cylindrical structure *(arrow)* that extends from the anterior end into the host cell (H). ×27,000. (From Aikawa et al., ref. 5, with permission.)

toward these erythrocytes (20). However, the invasion processes do not advance further. Electron microscopy shows the attachment between the apical end of a treated merozoite and rhesus erythrocyte, forming a junction similar to that seen for an untreated merozoite, but no further steps in the invasion process takes place. Several membrane-bound vacuoles with electron-translucent matrix appear in the erythrocyte cytoplasm near the attachment site.

When cytochalasin B-treated merozoites contact the erythrocyte membrane, freeze-fracture shows changes on the erythrocyte membrane similar to those seen during normal invasion with untreated merozoites (7). At the end of the invagination neck, IMPs on the P face of the erythrocyte membrane abruptly disappear so that the P face of the parasitophorous vacuole membrane possesses few IMP. Although the cytochalasin B-treated merozoites attach only to the erythrocyte membrane, the invagination of the erythrocyte extends far beyond the apical end of the merozoite (20) (Fig. 13). The significance of this finding is not yet clear.

The attachment between a cytochalasin-treated merozoite and a Duffy-positive human erythrocyte forms a junction. This attachment is identical to that seen for the rhesus erythrocyte. Again, no further invasion processes occur. In contrast, no junction is formed between a Duffy-negative human erythrocyte and cytochalasin B-treated merozoites. The apical end of the merozoite is oriented toward the erythrocyte, but instead of a junction, the erythrocyte is about 120 μm away from the merozoite and they are connected by thin filaments (20) (Fig. 14). Such filamentous attachments between cytochalasin B-treated merozoites and erythrocytes are not observed in the experiments with normal rhesus or Duffy-positive human eryth-

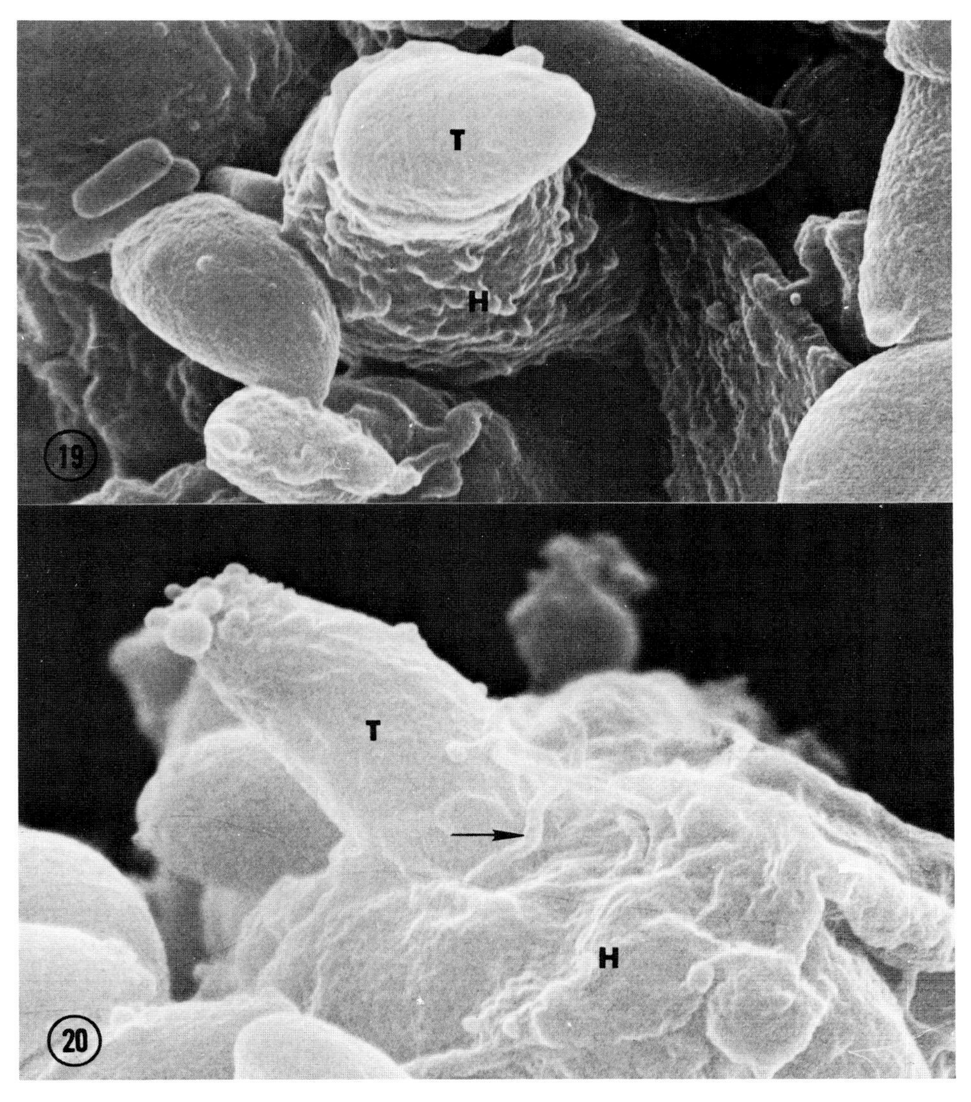

FIG. 19. Scanning electron micrograph of *T. gondii* (T) entering a host cell (H). ×19,000. (From Aikawa et al., ref. 5, with permission.)

Fig. 20. Scanning electron micrograph showing *T. gondii* (T) entering a host cell (H). Note pseudopods surrounding the parasite *(arrow)*. ×27,000. (From Aikawa et al., ref. 5, with permission.)

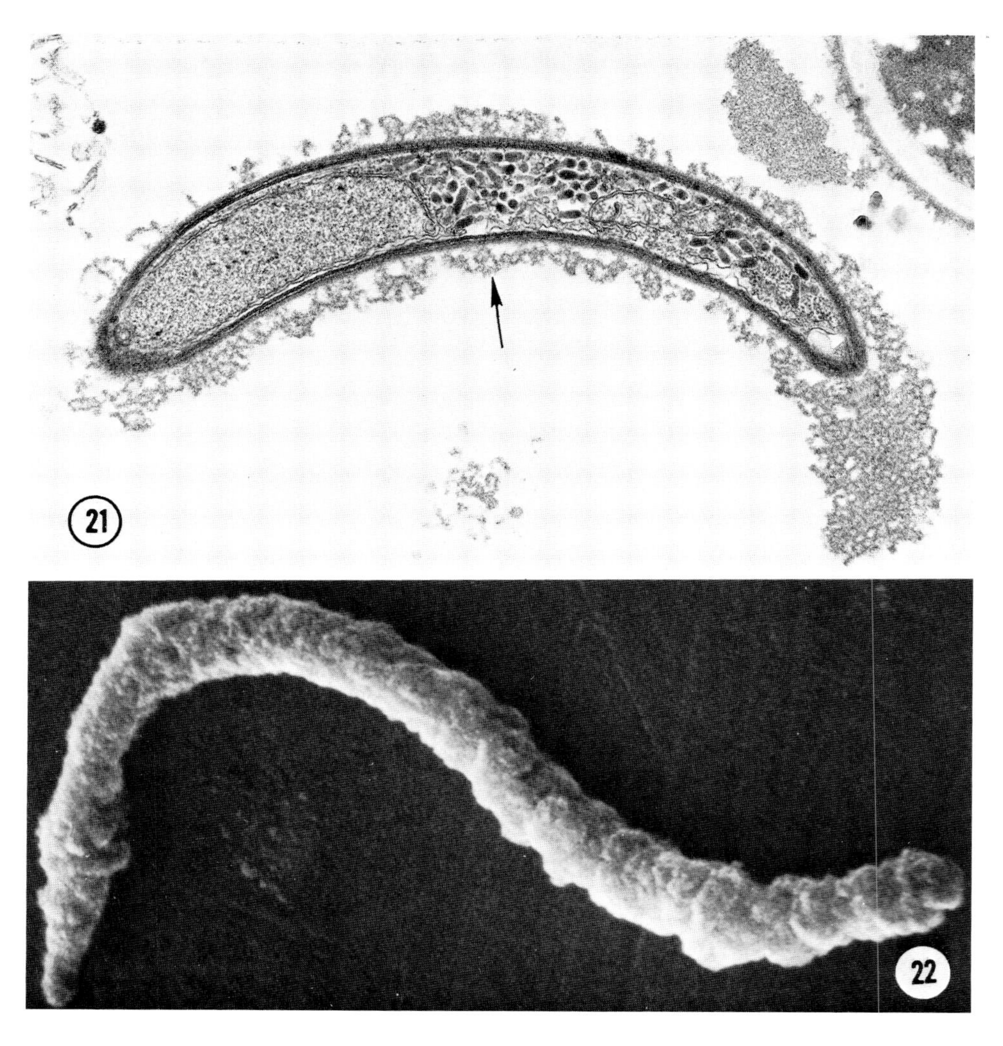

FIG. 21. A sporozoite incubated in immune serum is surrounded by a fibrillar coat *(arrow)*. ×15,000.
FIG. 22. Scanning electron micrograph of a sporozoite incubated in immune serum. The parasite displays an irregular surface configuration. ×14,000. (From Cochrane et al., ref. 11, with permission.)

rocytes. On the other hand, trypsinization of Duffy-negative human erythrocytes permits junction formation with cytochalasin B-treated merozoites.

These observations indicate that (a) cytochalasin B-treated merozoites attach to the host erythrocytes and form a junction, but cytochalasin B blocks the movement of this junction, preventing further invasion; and (b) the absence of junction for-

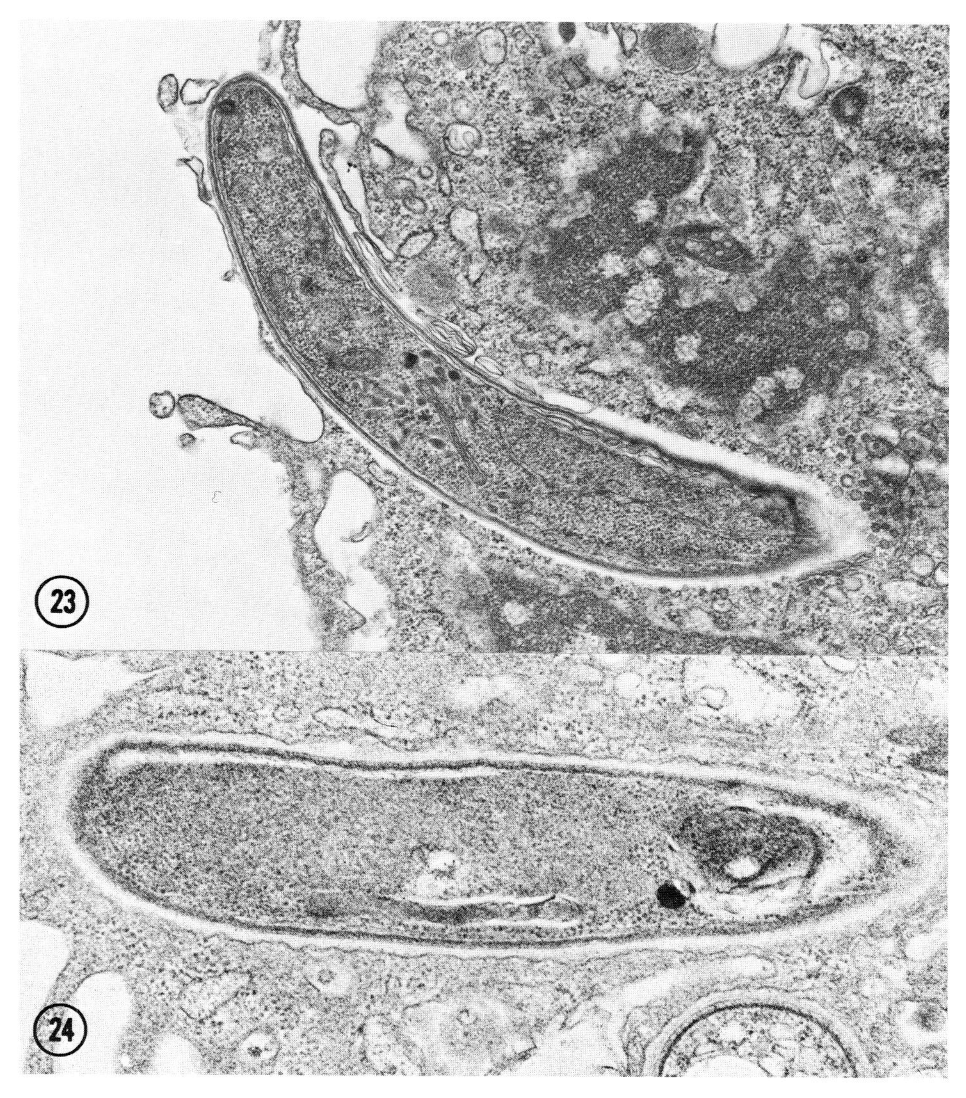

FIG. 23. Electron micrograph showing macrophage entry by a sporozoite. ×16,000. (From Danforth et al., ref. 12, with permission.)
FIG. 24. Sporozoite incubated with normal serum is situated within a vacuole of a macrophage. ×22,000.

mation with Duffy-negative human erythrocytes may indicate that the Duffy-associated antigen acts as a receptor for junction formation or a determinant on Duffy-negative erythrocytes blocks the junction formation.

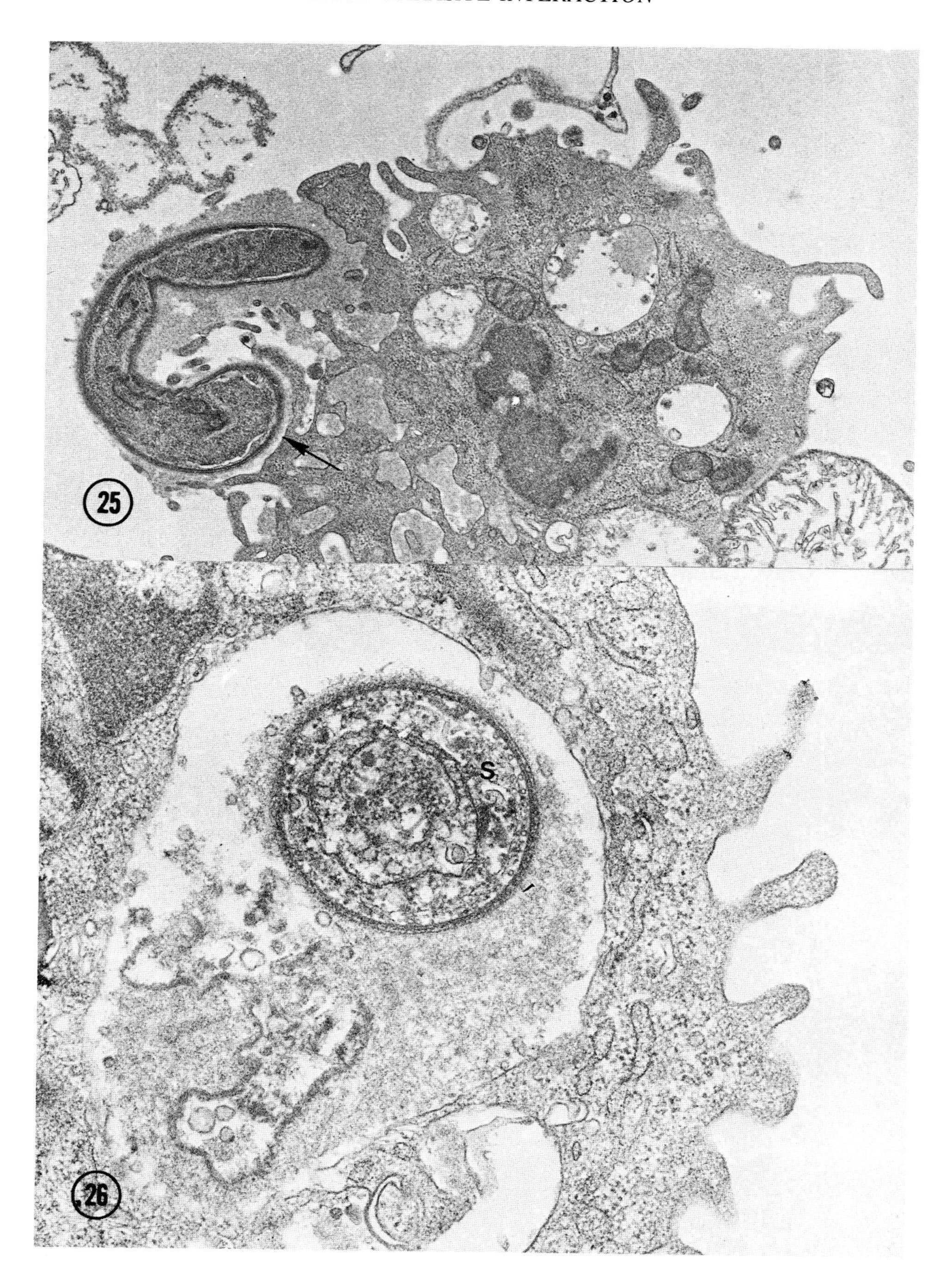
25
S
26

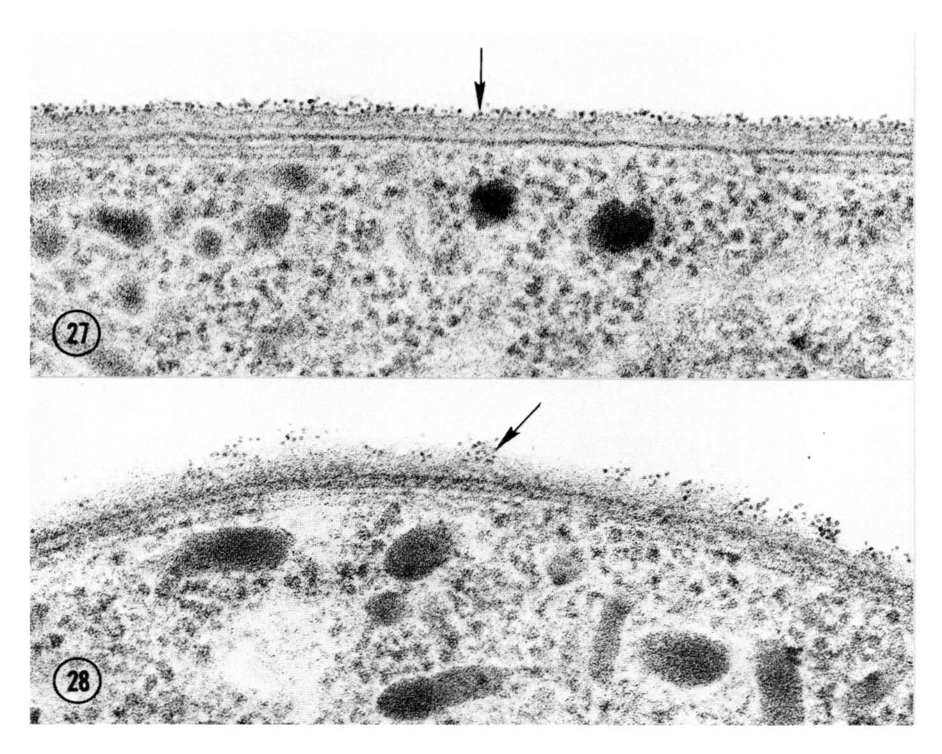

FIG. 27. Electron micrograph of a salivary gland sporozoite of *P. berghei* incubated with ferritin-conjugated monoclonal antibody showing a uniform distribution of ferritin particles *(arrow)* over the entire parasite surface membrane. ×80,000. (From Aikawa et al., ref. 8, with permission.)
FIG. 28. Electron micrograph of an oocyst-sporozoite incubated with ferritin-conjugated monoclonal antibody showing a patchy distribution of ferritin particles *(arrow)* on the surface. ×80,000. (From Aikawa et al., ref. 8, with permission.)

Interaction Between *T. gondii* and Macrophages

Toxoplasma gondii enters host cells by endocytosis. The tachyzoite of *T. gondii* is surrounded by a pellicle composed of two membranes and a row of subpellicular microtubules, similar to that of *Plasmodium* merozoites (5,25) (Fig. 15). A cyto-

FIG. 25. Sporozoite incubated with immune serum adheres to a macrophage *(arrow)*. The sporozoite is covered with a surface coat. ×13,000. (From Danforth et al., ref. 12, with permission.)
FIG. 26. Sporozoites (S) incubated with immune serum degenerate within the vacuole of the macrophage. ×31,000. (From Danforth et al., ref. 12, with permission.)

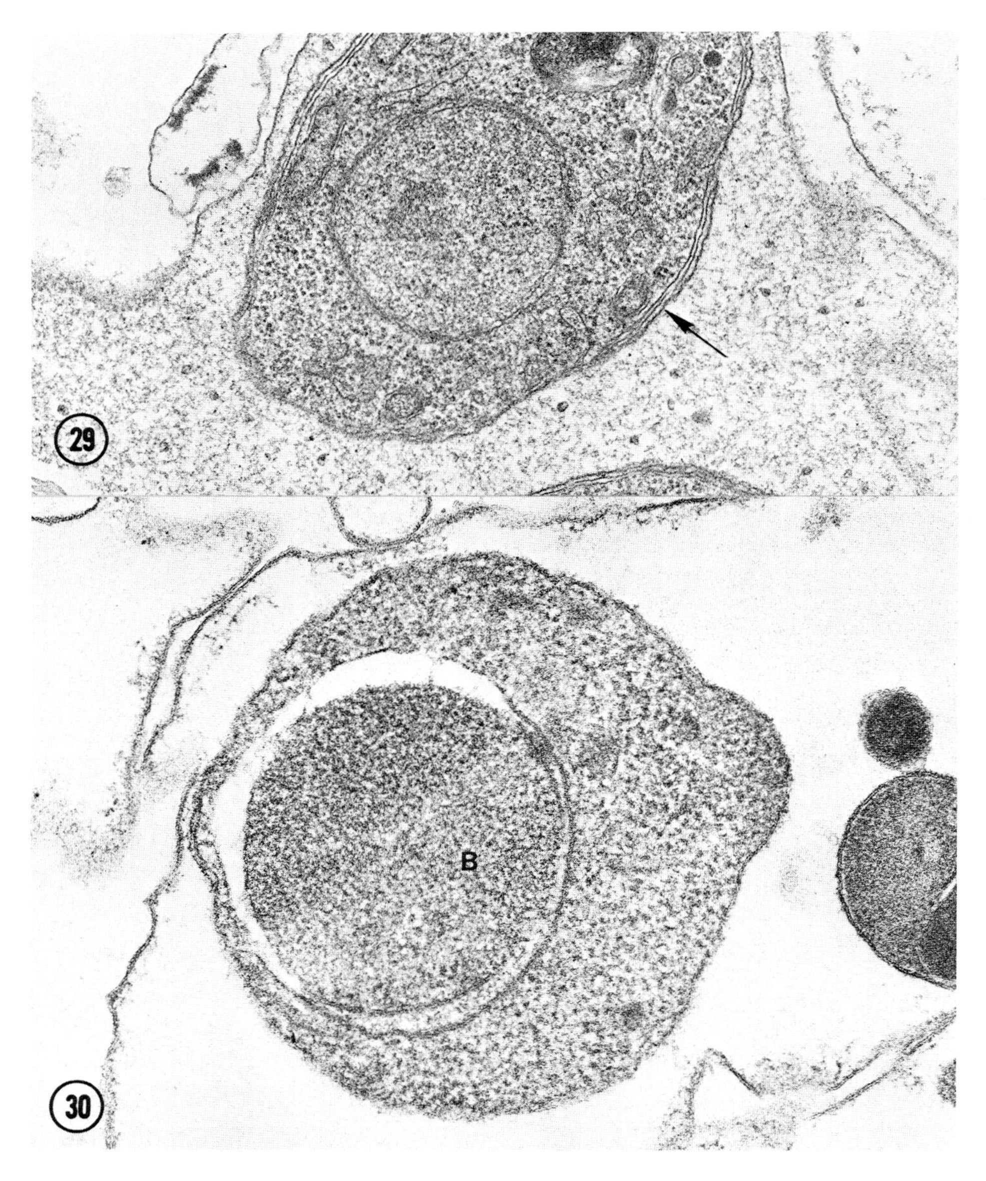

FIG. 29. *Babesia canis* within the erythrocyte. At this early stage, the parasite is surrounded by three membranes *(arrow)*. ×29,000.
FIG. 30. *Babesia canis* (B) within a lysed erythrocyte. No parasitophorous vacuole membrane surrounds the parasite. ×61,000.

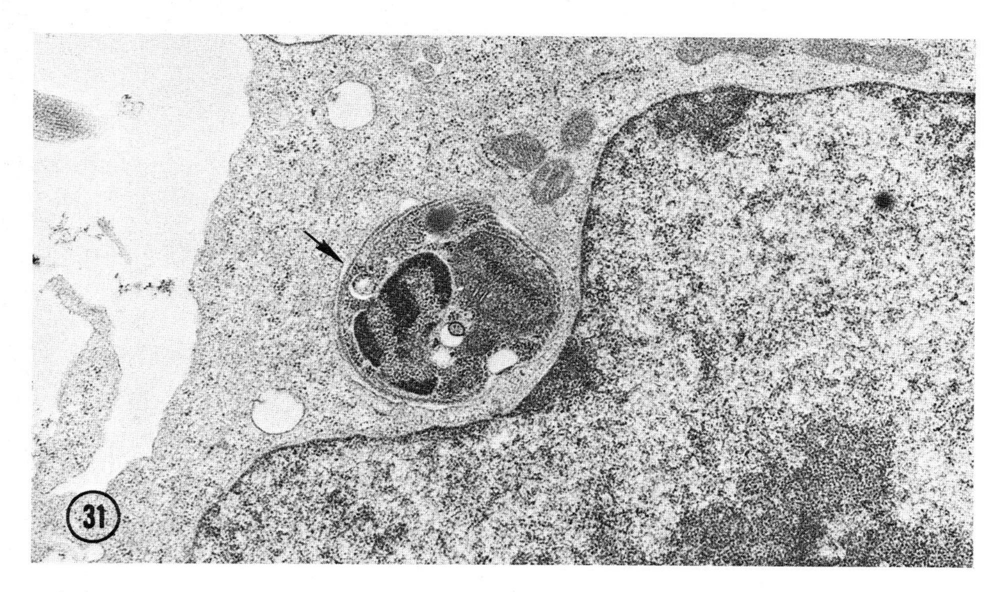

FIG. 31. An amastigotes of *Leishmania mexicana* within a macrophage. It is surrounded by the parasitophorous vacuole membrane *(arrow).* ×15,000.

stome is present in the pellicle. The anterior end is demarcated by electron-dense polar rings. Within the anterior end is a conoid—a hollow, truncated cone-shaped structure (Fig. 15). Electron-dense micronemes are present in the anterior and middle portions. Mitochondria are usually seen in the region anterior to the nucleus, while the nucleus is located in the middle portion.

Scanning electron microscopy shows the parasite to be crescent-shaped (Fig. 16). The anterior end is a truncated cone demarcated by a ring as seen in thin sections. The surface of the tachyzoite is finely granular, and slender ridges radiate from the anterior end to the posterior end. These ridges appear to correspond to the sub-pellicular microtubules. The cytostome is an indentation on the pellicle (Fig. 16). The nucleus is an elevated round structure in the midportion.

Transmission electron microscopy shows that the process of entry into host cells by *T. gondii* is initiated by the parasite contacting macrophages by its anterior end, which is oriented against the plasma membrane of the host cell, creating a small depression in the plasma membrane (Fig. 17). Scanning electron microscopy on the initial stages of the entry process shows a small depression on the surface of the host cell, into which the anterior end appears to be attached (5). A small portion of the host cell cytoplasm protrudes toward the anterior end of the parasite and seems to contact the anterior end or to extend inside it. Also, the parasites with a cylindrical structure are extending into the host cell cytoplasm (Fig. 18). This structure is continuous with the plasma membrane of the parasite and is surrounded by the plasmalemma of the host cell. No disruption of the host cell plasmalemma

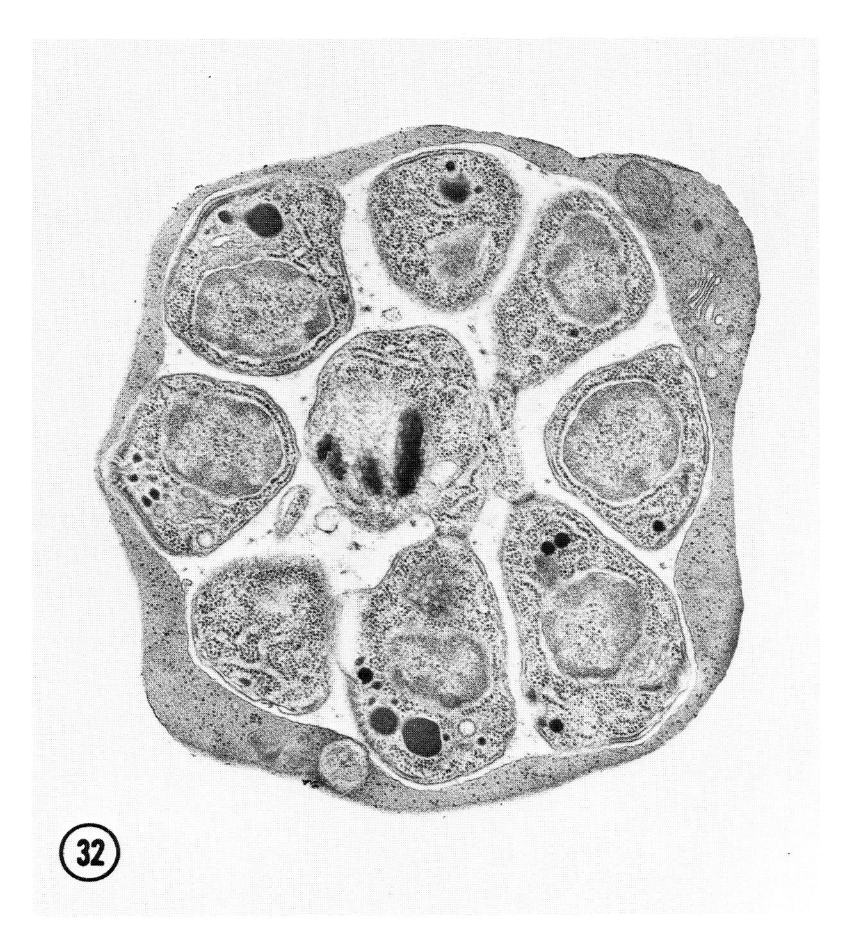

FIG. 32. Late stage of *P. cathemerium* development located within a parasitophorous vacuole. ×20,000. (From Aikawa, M., *Am. J. Trop. Med. Hyg.*, 15:449, 1966, with permission.)

is seen during this process. This structure appears to initiate and aid the invagination of the host cell membrane for parasite entry.

Rhoptries of *Toxoplasma* contain lysosomal enzymes that modify the host membrane and aid penetration by *Toxoplasma*. Subsequently, a slightly acidic protein with two components of molecular weights 70,000 and 150,000 have been isolated from tachyzoites of *T. gondii*. This protein enhances host cell infection both in cultured cells and in mice. Lycke et al. (18) suggested that this penetration-enhancing protein might alter the host membrane enzymatically and allow active penetration by the parasite.

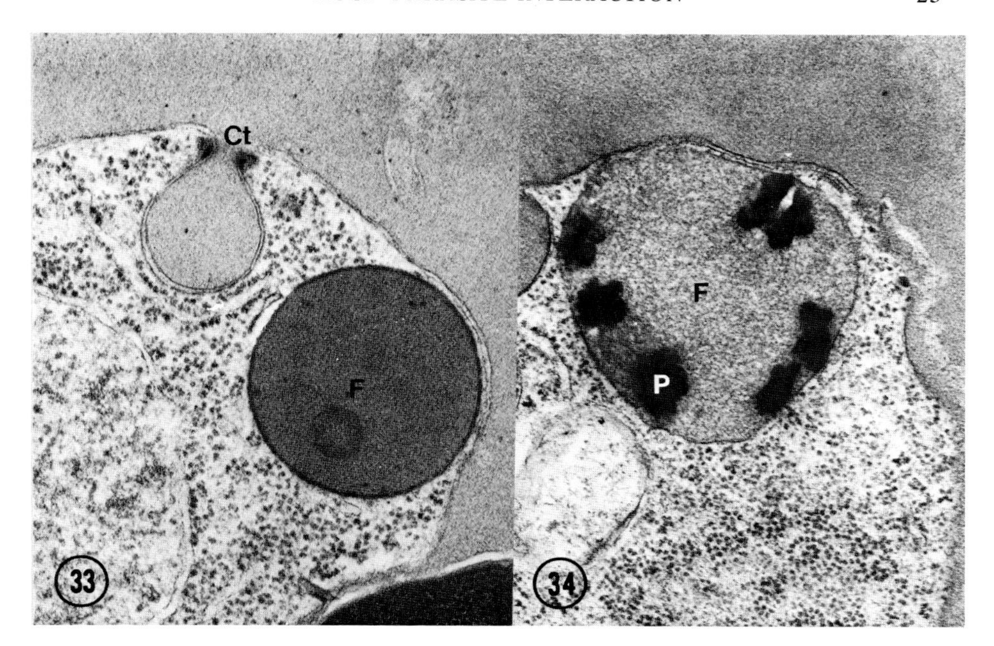

FIG. 33. A trophozoite of *P. gallinaceum* showing a cytostome (Ct) and food vacuole (F). The cytostome is ingesting host cell cytoplasm. ×32,000.
FIG. 34. A large food vacuole (F) containing malarial pigment particles (P). ×33,000.

As the parasite enters the host cell, pseudopods of the host cell become oriented toward the parasite and extend along the pellicle of *T. gondii*. Scanning electron microscopy demonstrates that these pseudopods surround the parasite and interdigitate with each other, so that the parasite becomes enmeshed (Figs. 19,20). After entry into the host cells, *T. gondii* is located in a parasitophorous vacuole as seen in *Plasmodium* (25). This observation again shows that *T. gondii* enters the host cells by endocytosis similar to that of *Plasmodium*. However, no junction has been demonstrated during host cell entry by *T. gondii*. The formation of the junction appears to occur only when protozoa interact with erythrocytes such as *Plasmodium* and *Babesia*.

Interaction Between Malarial Sporozoites and Macrophages

The interaction between malarial sporozoites and macrophages differs from that of the erythrocytic merozoites and erythrocytes.

It has been demonstrated previously that immune serum forms a thick surface coat on the surface of the sporozoites, while normal serum does not (11) (Figs. 21,22). When peritoneal macrophages and sporozoites treated with normal serum are incubated at 37°C for 1 hr, many sporozoites penetrated into the macrophages

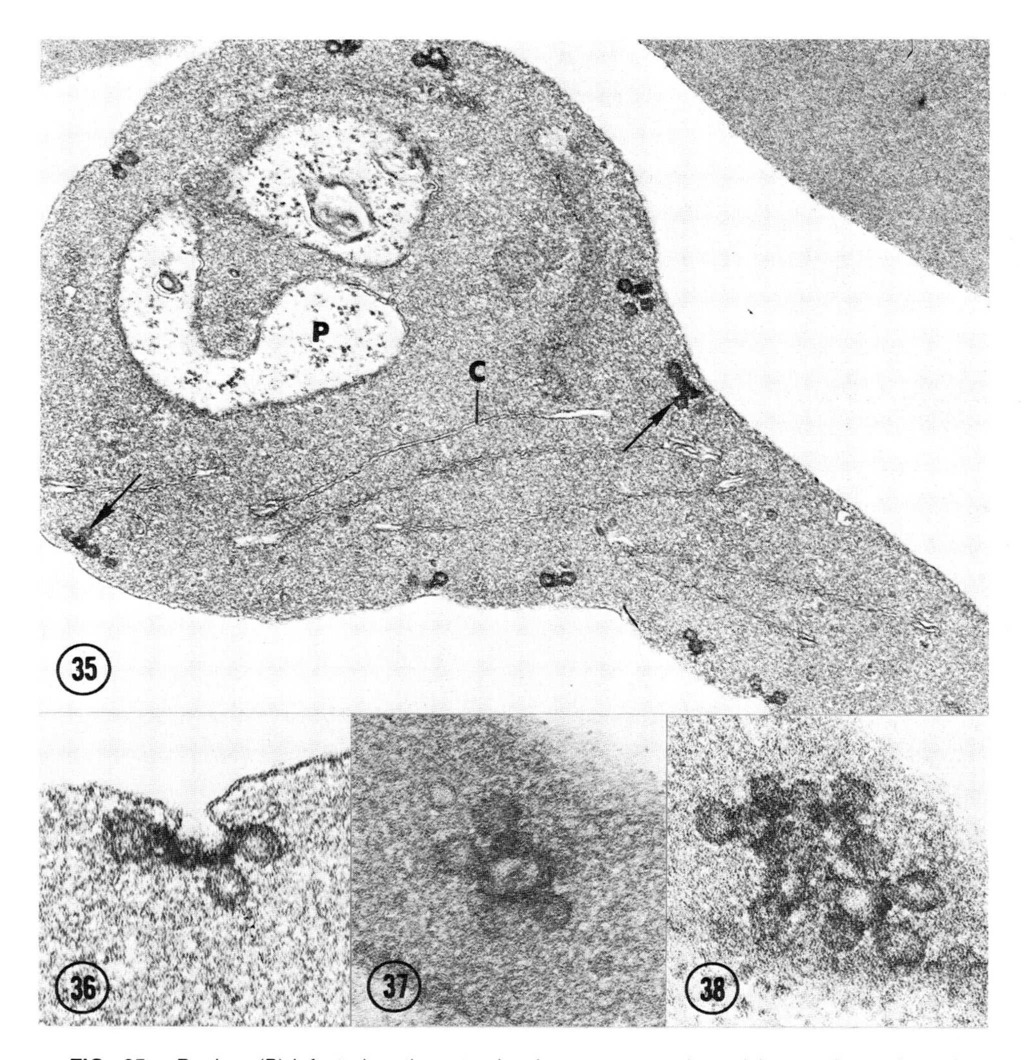

FIG. 35. *P. vivax* (P)-infected erythrocyte showing many caveola-vesicle complexes *(arrow)* along the membrane. Also seen are clefts (C) in the cytoplasm. ×31,000. (From Aikawa et al., ref. 4, with permission.)

FIG. 36.–38. High magnification of caveola-vesicle complexes in different configurations. ×97,000. (From Aikawa et al., ref. 4, with permission.)

and become intracellular (12) (Fig. 23). However, no degenerative changes are seen in these sporozoites, as shown in Fig. 24.

On the other hand, when peritoneal macrophages and sporozoites treated with immune serum are incubated together at 37°C, there are more intracellular sporo-

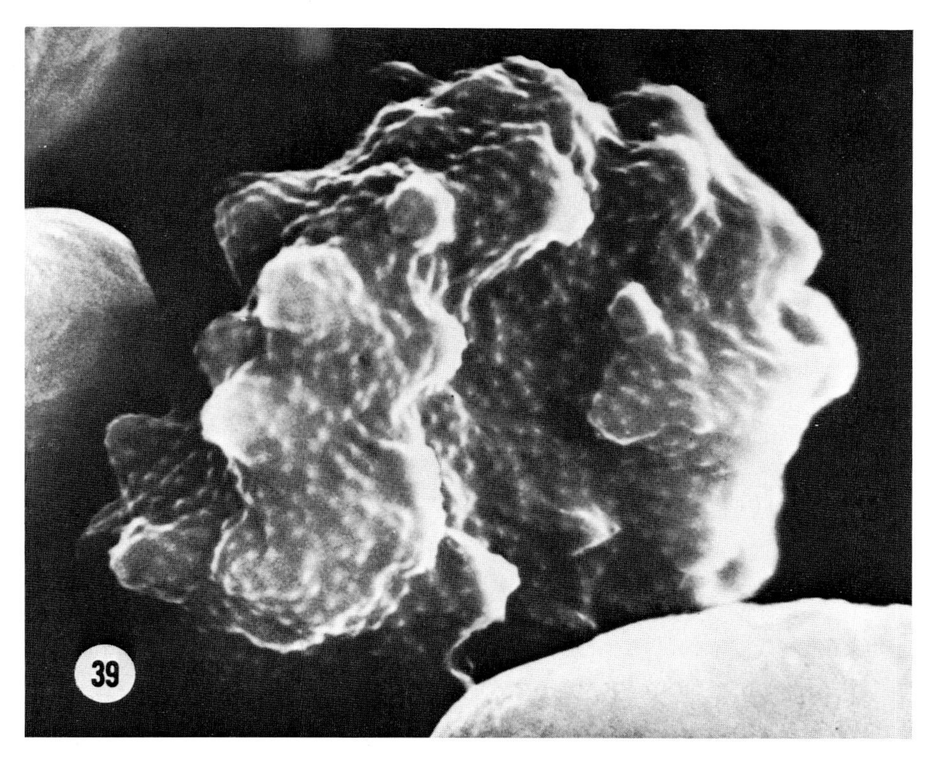

FIG. 39. Scanning electron micrograph of an erythrocyte infected with *P. falciparum*. Note numerous knobs on the surface. ×24,000. (From Aikawa, M. et al., *J. Parasitol., in press* 1982, with permission.)

zoites than that seen with sporozoites treated with normal serum. Sporozoites covered with a thick surface coat adhere to the macrophage (Fig. 25).

Intracellular sporozoites treated with immune serum are found within membrane-bound vacuoles and show degenerative changes (12). These sporozoites show a loss of ultrastructural organization and a breakdown of their pellicular membrane (Fig. 26). This observation differs from the interaction between the macrophages and sporozoites incubated with normal serum.

This experiment indicates that the presence or absence of immune serum on the sporozoite produces different interaction with macrophages. The surface coat covering immune-serum-treated sporozoites may be responsible for the macrophage to digest the sporozoites. On the other hand, sporozoites treated with normal serum may survive within the macrophages. Since it has been suggested that sporozoites gain entry into the hepatocytes through the Kupffer cells, to become exoerythrocytic stages, these intact sporozoites demonstrated here may be responsible for the formation of the exoerythrocytic stages in the liver.

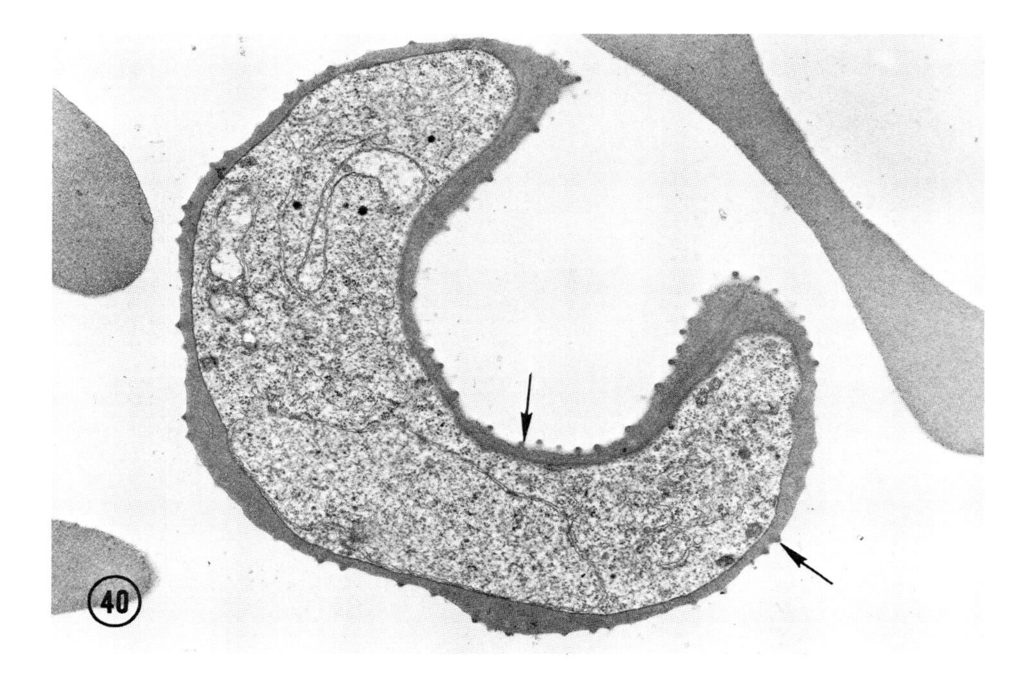

FIG. 40. Transmission electron micrograph of an erythrocyte infected with *P. falciparum*. Note knobs *(arrow)* along the plasma membrane. ×27,000.

Interaction Between Malarial Sporozoites and Monoclonal Antibody

Recently, it has been reported that a hybridoma, found by fusion of a plasmacytoma cell line with spleen cells of mice immunized with irradiated *P. berghei* salivary gland sporozoites, produced a large amount of monoclonal antibodies against a surface antigen (Pb44) of this parasite (29). Passive transfer of small doses of purified antibody completely protects against a sporozoite-induced infection. The presence of Pb44 on the *P. berghei* membrane is demonstrated by using this monoclonal antibody labeled with ferritin particles (8). The monoclonal antibody produces a surface coat over the sporozoite. Salivary gland sporozoites of *P. berghei*, which are incubated at 0°C with the ferritin conjugate, shows a uniform distribution of ferritin particles over the entire parasite membrane (Fig. 27). In contrast, sporozoites of oocysts show either a complete absence of ferritin or a patchy distribution of ferritin particles on their surface (Fig. 28). Using the same ferritin-conjugated preparation of monoclonal antibody, no label is found on the erythrocytic stages.

To determine the stage at which Pb44 disappear from the subsequent developmental stages of *P. berghei*, exoerythrocytic forms of the parasite were examined. Twelve, 18, and 24 hr after the inoculation of sporozoites, the exoerythrocytic forms are detected both by light microscopy and by immunofluorescence using monoclonal antibodies. However, after 30 hr, the exoerythrocytic forms are *only* seen by light microscopy and *not* by immunofluorescence (8). This finding indicates

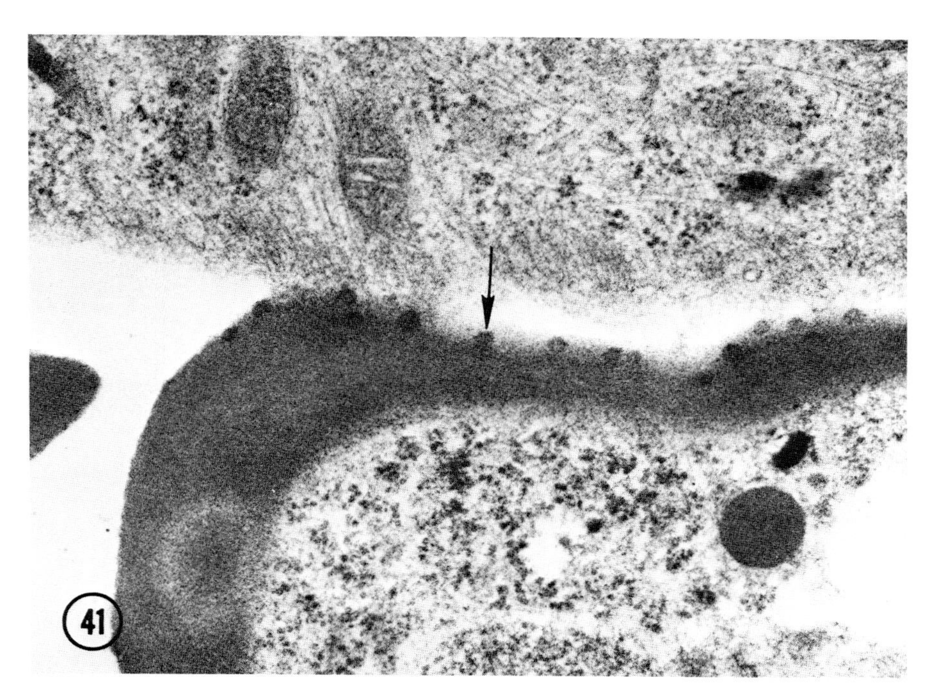

FIG. 41. A *P. falciparum* infected erythrocyte attaches to an endothelial cell via knobs *(arrow)*. Note the aggregations of knobs on the erythrocyte membrane adjacent to the endothelial cells. ×56,000. (From Udeinya et al., ref. 27, with permission.)

that the membrane-associated Pb44 disappear after sporozoites penetrate the hepatocytes. It supports the idea that Pb44 is a differentiation antigen, involved in the unique, and essential function associated with salivary gland sporozoites.

Therefore, it is possible that Pb44 plays an important role in the process of recognition and penetration of the parasite into the target host cells (hepatocytes). If this is indeed the case, the search of homologs of Pb44 in the sporozoites of human malaria would be of considerable interest, since these molecules would be ideally suited for the preparation of vaccines.

GROWTH OF INTRACELLULAR PROTOZOA IN HOST CELLS

After completion of host cell entry by protozoa, the parasite is surrounded by a parasitophorous vacuole membrane. The question of resistance is especially pertinent for protozoa such as *Trypanosoma*, *Toxoplasma*, and *Leishmania* that infect the phagocytic cells and establish resistance within the phagosome. Several lines of evidence indicate that soon after the completion of endocytosis within phagocytic as well as nonphagocytic cells, the nature of the parasitophorous vacuole membrane is altered (1). This may be one of the main determinants in assuring a safe and suitable intracellular environment for the parasite.

Protozoa that parasitize macrophages do not share a common defense mechanism to escape digestion within a phagosome. In *Trypanosoma cruzi*, the problem seems to be solved by the disintegration of the vacuole membrane. Trypomastigotes were shown to remain within the phagocytic vacuoles for 60 to 90 min, but at later periods they were found free in the cytoplasmic matrix (22). Similarly, the parasitophorous vacuole membrane surrounding *Babesia* disappears and *Babesia* continues development in the erythrocyte cytosol (23), although invasion of *Babesia* initially resembles that of *Plasmodium* (Figs. 29 and 30). In *Toxoplasma*, even though the parasites remain within the vacuole, the vacuole membrane appears different from that of a regular phagosome and usually does not fuse with lysosomes (14). Similarly, amastigotes of *Leishmania* remain within the parasitophorous vacuoles (Fig. 31). Even though lysosomes of the macrophage fuse with the parasitophorous vacuole, *Leishmania* survive and multiply (9). This again indicates the alteration of the vacuole membrane from the host plasma membrane.

In malarial infections, the vacuole membrane grows with the developing parasite (Fig. 32) and is retained until formation of the next generation of merozoites. Changes in the molecular organization of this inside-out erythrocyte membrane are apparent very early in the development. Freeze-fracture studies have shown major differences in the distribution of intramembranous particles of the erythrocytic and vacuolar membranes (16). In addition, cytochemical studies indicated differences in surface change, glycoprotein, and enzyme distribution between these two membranes (17).

Once the parasite is successfully established within a host cell, it grows, multiples, and eventually destroys the host. Intracellular protozoa rely on the host cells as their major source of nutrients. While some nutrients may be acquired by the parasite by diffusion or transport through the surrounding membranes, others are obtained by ingestion of host cell cytoplasm through a specific organelle, the cytostome (3). The cytostome is a circular structure present in the pellicle. In *Plasmodium* the host cell cytoplasm enters the cytostomal cavity (Fig. 33), from which small vacuoles are pinched off. After ingestion, the host cell cytoplasm is surrounded by a unit membrane and becomes a food vacuole, equivalent to a phagosome (3). Electron-dense pigment, hemozoin, is formed within the food vacuoles as digestion proceeds. As hemozoin is formed, there is a concomitant decrease in the density of the food vacuole content (Fig. 34). Cytochemical studies have shown acid phosphatase activity within the vacuoles (2). Based on these studies, it can be concluded that *Plasmodium* possesses a lysosomal system of its own that digests host cell cytoplasm for its nutrients.

HOST CELL ALTERATION

Intracellular *Plasmodium* not only influences the nature of the parasitophorous vacuole membrane, but also the host plasma membrane. The structures that occur in infected erythrocytes and that can be seen by light microscopy have been given various names such as Schüffner's dots, Mauer's clefts, Zieman's stipplings, etc.

Electron microscopy shows that two types of erythrocyte membrane modifications have been observed in infections with certain species of *Plasmodium:* (a) electron-dense protrusions that have been called knobs and (b) caveola-vesicle complexes along the erythrocyte membrane.

Schüffner's dots seen in the erythrocytes infected by vivax-type and ovale-type parasites were demonstrated to be caveola–vesicle complexes along the erythrocyte plasma membrane (4). They consist of caveolae surrounded by a small vesicle arranged in an alveolar fashion (Figs 35–38). Horseradish-peroxidase-labeled immunoglobulin from monkeys infected with *P. vivax* binds to the vesicle membrane (4), indicating the presence of marlarial antigens within them. After incubation of viable parasitized erythrocytes with ferritin, ferritin particles appear within the vesicles, indicating that these vesicles are pinocytotic in origin. Maurer's clefts seen by light microscopy appear to correspond to narrow slit-like structures in the cytoplasm that could be extrusions of the parasitophorous vacuole membrane (4). Another prominent change in infected erythrocytes is the development of knobs on the erythrocyte membrane (Figs. 39,40). The knobs appear mostly on the erythrocytes infected with falciparum-type parasites (4,16). The knobs form focal junctions with the endothelial cell membranes, suggesting that they are responsible for the sequestration of infected erythrocytes in deep organs.

Recently an *in vitro* study (27) demonstrated that erythrocytes infected with *P. falciparum* bind specifically to human endothelial cells in culture and that the knobs on the infected erythrocyte membrane are the points of attachment to the endothelial cells (Fig. 41). Knobs were concentrated in the area of the erythrocyte membrane in apposition with the endothelial cells. However, serum from an immune monkey containing antibodies directed against the knobs completely abolish binding. In ovale-type malaria, the erythrocyte membrane shows both knobs and caveolar-vesicle complexes.

SUMMARY

The interaction between the host cells and intracellular protozoa, with particular emphasis on *Plasmodium*, has been reviewed. Three major aspects basic to host–parasite interaction are (a) host cell entry by intracellular protozoa, (b) growth of intracellular protozoa within the host cells, and (c) alteration of the host cells.

All intracellular protozoa enter into the host cell by endocytosis with the exception of *Nosema*. However, little is known of the mechanisms and factors that trigger the initiation of endocytosis. On the other hand, some experimental data indicate the possible presence of a specific receptor for host cell entry by protozoa. A distinct junction appears between the erythrocyte membrane and the parasite during the endocytosis process in the case of *Plasmodium* and *Babesia*. The movement of this junction during entry may be an important component of the endocytotic mechanisms by which *Plasmodium* and *Babesia* enter the host cell.

After completion of host cell entry, a parasitophorous vacuole membrane surrounds the intracellular protozoa. Even though the vacuole membrane originates from the host cell membrane, intracellular protozoa alter the nature of this host-derived membrane. This alteration of the derived membrane may be one of the main determinants in assuring a safe and suitable intracellular environment for the intracellular protozoa.

Once protozoa are successfully established within a host cell, they grow and multiply. These parasites rely on the host cells as a main source of nutrients. After ingestion of host cell cytoplasm, this material becomes incorporated into a food vacuole, in which it is digested by various enzymes, including acid phosphatase.

Our knowledge of specific mechanisms involving interaction between intracellular protozoa and host cells is still fragmentary. Difficulty in studying intracellular protozoa comes mainly from the difficulty of complete isolation of the parasites from the host and precise control of experimental conditions. The advent in recent years of culturing some protozoa has provided clean protozoa free from host contamination. This advance may facilitate our understanding of intracellular protozoa in relation to the host.

ACKNOWLEDGMENTS

This study was in part supported by grants from the U.S. Public Health Service (AI-10645 & AI-16680), the World Health Organization (T16/181/M2/52), and The U.S. Army R & D Command (DAMD 17-79-C-9029).

REFERENCES

1. Aikawa, M., and Kilejian, A. (1979): Invasion procedures and intracellular localization of parasitic protozoa. In: *Lysosomes*, Vol. 6, edited by J. T. Dingle, P. J. Jacques, and I. H. Shaw, North-Holland, Amsterdam, pp. 31–48.
2. Aikawa, M., and Thompson, P. E. (1971): Localization of acid phosphatase activity in *Plasmodium berghei* and *Plasmodium gallinaceum:* An electron microscopic observation. *J. Parasitol.*, 57:603–610.
3. Aikawa, M., Hepler, P. K., Huff, C. G., and Sprinz, H. (1966): Feeding mechanism of avian malarial parasites. *J. Cell. Biol.*, 28:355–373.
4. Aikawa, M., Miller, L. H., and Rabbege, J. (1975): Caveola-vesicle complexes in the plasmalemma of erythrocytes infected with *Plasmodium vivax* and *Plasmodium cynomalgi:* Unique structures related to Schüffner's dots. *Am. J. Pathol.*, 79:285–300.
5. Aikawa, M., Komata, Y., Asai, T., and Midorikawa, O. (1977): Transmission and scanning electron microscopy of host cell by *Toxoplasma gondii. Am. J. Pathol.*, 87:285–290.
6. Aikawa, M., Miller, L. H., Johnson, J., and Rabbege, J. R. (1978): Erythrocyte entry by malarial parasites: A moving junction between erythrocyte and parasite. *J. Cell. Biol.*, 77:72–82.
7. Aikawa, M., Miller, L. H., Rabbege, J. R., and Epstein, N. (1981): Freeze-fracture study on the erythrocyte membrane during melarial parasite invasion. *J. Cell. Biol. (in press).*
8. Aikawa, M., Yoshida, N., Nussenzweig, R. S., and Nussenzweig, V. (1981): The protective antigen of malarial sporozoites *(Plasmodium berghei)* is a differentiation antigen. *J. Immunol.*, 126:2494–2495.
9. Chang, K.-P., and Dwyer, D. H. (1978): *Leishmania donavani:* Hamster macrophage interactions *in vitro*: Cell entry, intracellular survival and multiplication of amastigotes. *J. Exp. Med.*, 147:515–530.
10. Chapman, W. E., and Ward, P. A. (1977): *Babesia rodhaini:* Requirement of complement for penetration of human erythrocytes. *Science*, 196:67–70.
11. Cochrane, A. H., Aikawa, M., Jeng, M., and Nussenzweig, R. S. (1976): Antibody-induced ultrastructural changes of malarial sporozoites. *J. Immunol.*, 116:859–867.

12. Danforth, H. D., Aikawa, M., Cochrane, A. H., and Nussenzweig, R. S. (1980): Sporozoites of mammalian malaria: Attachment, interiorization and intracellular fate within macrophages. *J. Protozool.*, 27:193–201.
13. Griffin, F. M., Griffin, J. A., Leider, J. E., and Silverstein, S. C. (1975): Studies on the mechanism of phagocytosis. I. Requirements for circumferential attachment of particle-bound ligands to specific receptors on the macrophage plasma membrane. *J. Exp. Med.*, 142:1263–1282.
14. Jones, T. C., and Hirsch, J. G. (1972): The interaction between *Toxoplasma gondii* and mammalian cells. II. The absence of lysosomal fusion with phagocytic vacuoles containing living parasites. *J. Exp. Med.*, 136:1173–1194.
15. Kilejian, A. (1976): Does a histidine-rich protein from *Plasmodium lophurae* have a function in merozoite penetration? *J. Protozool.*, 23:272–277.
16. Kilejian, A., Abati, A., and Trager, W. (1977): *Plasmodium falciparum* and *Plasmodium coatneyi:* Immunogenecity of "knob-like protrusions" on infected erythrocyte membranes. *Exp. Parasitol.*, 42:157–164.
17. Langreth, S. G. (1977): Electron microscope cytochemistry of host-parasite membrane interactions with malaria. *WHO Bull.*, 55:171–178.
18. Lycke, E., Carlberg, K., and Norrby, R. (1975): Interactions between *Toxoplasma gondii* and its host cells: Function of the penetration-enhancing factor of *Toxoplasma. Infect. Immun.*, 11:853–861.
19. Miller, L. H., Haynes, J. D., McAuliffe, F. M., Shiroishi, T., Durocher, J. R., and McGinniss, M. H. (1977): Evidence for differences in erythrocyte receptors for malarial merozoites. *J. Exp. Med.*, 146:277–281.
20. Miller, L. H., Aikawa, M., Johnson, J. G., and Shiroishi, T. (1979): Interaction between cytochalasin B-treated malarial parasites and red cells: Attachment and junction formation. *J. Exp. Med.*, 149:172–184.
21. Nicolson, G. L., Marchesi, V. T., and Singer, S. J. (1971): The localization of spectrin on the inner surface of human red blood cell membranes by ferritin conjugated antibodies. *J. Cell. Biol.*, 51:265–272.
22. Nogueira, N., and Cohn, Z. (1976): *Trypanosoma cruzi:* Mechanisms of entry and intracellular fate in mammalian cells. *J. Exp. Med.*, 143:1402–1420.
23. Rudzinska, M., Trager, W., Lewengrub, J., and Gubert, E. (1976): An electron microscope study of *Babesia microti* invading erythrocytes. *Cell Tiss. Res.*, 169:323–334.
24. Seed, T. M., Aikawa, M., Sterling, C. R., and Rabbege, J. (1974): Surface properties of extracellular malaria parasites: A morphological and cytochemical study. *Infect. Immun.*, 9:750–761.
25. Sheffield, H. G., and Melton, M. L. (1968): The fine structure and reproduction of *Toxoplasma gondii. J. Parasitol.*, 54:209–226.
26. Timm, A. H. (1978): The ultrastructural organization of the contractile peripheral protein layer of the human erythrocyte membrane. *J. Anat.*, 127:415–424.
27. Udeinya, I. J., Schmidt, J. A., Aikawa, M., Miller, L. H., and Green, L. (1981): *Falciparum* malaria-infected erythrocytes specifically bind to cultured human endothelial cells. *Science*, 213:555–557.
28. Weidner, E. (1976): The microsporidian spore invasion tube: The ultrastructure, isolation, characterization of protein comprising the tube. *J. Cell Biol.*, 71:23–34.
29. Yoshida, N., Nussenzweig, R. S., Potocnjak, P., Nussenzweig, V., and Aikawa, M. (1980): Hybridoma produces protective antibodies directed against the sporozoite stage of malarial parasite. *Science*, 207:71–73.

Molecular Biology of Parasites, edited by
J. Guardiola, L. Luzzatto, and W. Trager.
Raven Press, New York © 1983.

Recognition and Invasion by Intracellular Parasites

Milton J. Friedman

Cancer Research Institute, University of California, San Francisco, California 94143

Parasites and their hosts are complex organisms, and we cannot hope to completely describe their effects on each other. Instead, our goal should be to identify the critical determinants of their interactions, and the critical determinants of infectious diseases are those that influence who gets sick and who dies. For example, morbidity and death from falciparum malaria are determined by the parasite multiplication rate, the ability of the body to withstand parasitization, and the timing of the protective response (Fig. 1). The factors that influence these variables are of great importance.

The problem we have is to identify those factors, to measure them, and then hopefully to alter the course of disease. Two methods of identification are discussed by Luzzatto and Lucas in this volume. They are the correlation of genetic background and environmental factors with disease occurrence. Both of these methods can be included in the epidemiological sciences. Another approach might be called biochemical epidemiology—the correlation of biochemical factors with incidence or severity of infection.

The geographical distribution of falciparum malaria is determined by man-mosquito contact. Where contact is sufficient to support the cycle of infection, the parasite may be found. If there were a nonhuman mammalian host of *Plasmodium falciparum*, the situation would be more complex, but there is none in nature. Thus, the host specificity of this parasite is a primary determinant of disease occurrence. But what are the factors that determine infection? What are the factors that allow a parasite to recognize and infect a competent host?

Studies on the ability of intracellular parasites to recognize and invade their host cells have failed to identify the specific host surface molecules to which parasites

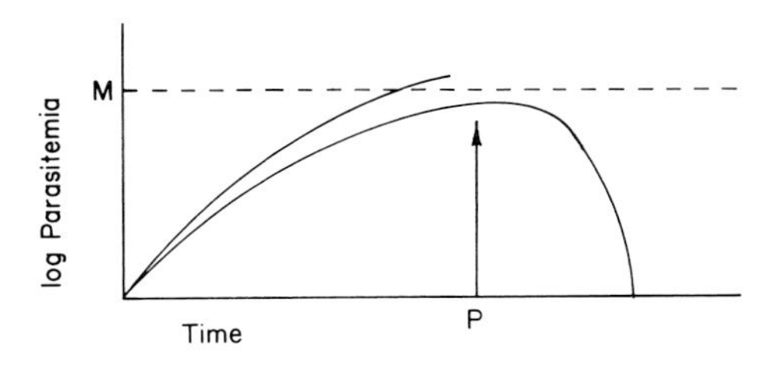

FIG. 1. The critical factors in falciparum malaria can be related to the three parameters shown in this graph: (1) The rate of parasite increase, y/x; (2) the mortality threshold, M; and (3) the time of a protective response, P. The physiological and biochemical determinants of these factors have yet to be identified. All we know is that in many cases y, the parasitemia, exceeds M and death results.

bind and to precisely describe the biochemical events of infection. This chapter presents a general and methodological analysis of the problem. For convenience, I have separated the events of recognition and invasion, but the parasite may not make this distinction.

Recognition can be a result of chemotaxis, which requires the binding of soluble ligands, and of cell-cell association, which requires the binding of surface ligands. The binding of soluble ligands by cell surfaces resembles enzyme-substrate binding. A binding curve can be generated that is characterized by a binding constant equal to the concentration of ligand in which 50% of the receptors are occupied. The concentration giving 50% saturation is a critical point, the point at which response is most likely to occur. This response must be mediated by a transducing mechanism that will in some degree influence the stage of all the receptors. Coupling of receptors can cause positive or negative cooperativity and an increase in the complexity of the data. An additional level of complexity arises with multivalent ligands.

Cellular responses to multivalent ligands are both dramatic and complex. Typically, the binding affinity of a multivalent ligand is three to four orders of magnitude greater than that of the monovalent ligand. Furthermore, cellular responses to multivalent ligands are highly dependent on receptor mobilities. If receptor mobility is zero as at low temperature, the binding affinity of the multivalent ligand may be the same as that of the monomer. When temperature is increased and diffusion is allowed, the binding affinity then increases many fold. At critical ligand concentrations, this effect can operate as a cellular switch, making the cell either responsive or refractory to signal transduction. The effect can be modulated by the cytoskeleton of the cell.

This same sensitivity to multivalent ligands becomes central to surface-surface interactions and parasite-host cell binding. Here, however, we are dealing with two modulating surfaces, and the possible interactions are both more numerous and more complex. It is helpful to look at actual binding studies that have been done with liposome, virus, and protozoan binding to cells.

McConnell's group at Stanford has constructed antigen-bearing liposomes with either fluid or solid lipid phases (8). They can coat the liposomes with varying densities of antibody and measure binding and phagocytosis by macrophages. When antibody density is high on the liposome surface, there is no difference in binding for fluid or solid vesicles; but when antibody density is low, solid vesicles bind very weakly, whereas fluid vesicles bind well. Apparently, the slowness of diffusion in solid vesicles prevents the formation of the multivalent binding sites necessary to hold vesicle and cell together in a turbulent environment.

Virus binding is strongly dependent on temperature, and Noble-Harvey and Lonberg-Holm (12) suggest that in this case, receptor mobility on the host cells determines the ability to form multivalent binding sites. Furthermore, metabolic inhibitors decrease rhinovirus binding to HeLa cells, and thus, active modulation of receptor distribution may also play a role in forming the multivalent site (9).

Both of these phenomena are critical to *Leishmania tropica* attachment to macrophages. Promastigote binding was inhibited by low temperature, mild fixation of

properties are cytochalasin-sensitive, but it is clear that the parasite is the active partner in invasion. Studies on other parasites have shown different results. Thus infection of macrophages by *Leishmania* amastigotes appears to procede by host-mediated endocytosis. Normal phagocytosis would deliver amastigotes into the lysosomal system and in fact, vacuoles containing parasites fuse with secondary lysosomes (3). The survival of the parasites in macrophages is still unexplained.

When *Toxoplasma* infects the macrophage, it too enters an endocytic vacuole, but this vacuole does not fuse with secondary lysosomes. Jones and Hirsch (6) have suggested that fusion is actively inhibited by the parasite, but it is equally possible that the parasite, by being the active partner, does not enter a phagosome at all, but merely a parasite-formed vacuole. Only 50% of toxoplasma survive the infection; the others are killed by lysosomal fusion. These may have entered by the macrophage-active route. That toxoplasma can be the active partner is demonstrated by its infection of fibroblasts and HeLa cells (7).

We have seen that some parasites, including SFV and *Leishmania*, are phagocytized and then escape from the normal degradative process. Others like *Toxoplasma*, *Eimeria*, and *Chlamydia* stimulate their own uptake into privileged cellular sites. It is likely that the discriminating events in these processes take place at the cell surface, in how and to what the parasite binds. So until we are able to identify specific host binding sites, we will be unable to understand these events fully. In the malaria system, we have made some progress in the past year.

Merozoites of *P. falciparum* invade neuraminidase-treated human red blood cells (N-RBC) in human serum but not baboon serum. This activity is absent if human serum is pretreated with antiserum to orosomucoid (OR; also α_1-acid glycoprotein). OR is a serum protein with complex oligosaccharides containing sialic acid. It binds to N-RBC and restores a negatively charged surface to the red cell. In other respects, it does not resemble the major red cell surface proteins, suggesting that surface charge is a primary determinant of host cell competence for infection. Baboon red cells are infected little, if at all, in human serum, but merozoites do bind to these cells. (They do not bind in baboon serum.) Clearly, human cells have a further determinant of infection, one not found in the baboon cell.

The ability of OR to mediate attachment to N-RBC is not required by normal human red cells, and indeed, OR will inhibit merozoite entry at higher concentrations. The inhibitory activity is proportional to OR concentration, increasing in the physiological range. An acute phase reactant, OR increases during malaria infection from 0.7 to 2.5 mg/ml. Inhibition *in vitro* with these concentrations is approximately 25 and 80%, respectively. Thus, the OR acute-phase response is protective and may be critical in preventing lethal levels of parasitemia in natural infection.

Infection of N-RBC in the presence of OR can be prevented by treatment with trypsin. We believe that trypsin removes the second determinant, but this is not yet proven. This determinant is not glycophorin A, as removal of this protein is not sufficient to prevent infection. We are currently attempting to identify the second determinant.

We hope in these studies to eventually account for the ability of the falciparum malaria parasite to both recognize the human red blood cell and to invade this otherwise nonendocytic cell. These abilities, as discussed earlier, are the primary determinants of where, when, and in whom illness and death from malaria occur. OR and other substances that inhibit invasion (for example, merozoite-specific antibodies) are also critical determinants of malaria's effects. The ability of OR to decrease the slope of the curve in Fig. 1 could have a profound effect on mortality. Studies of this kind in other parasite host systems may be equally revealing.

ACKNOWLEDGMENTS

This investigation received the financial support of the UNDP/World Bank/WHO Special Programme for Research and Training in Tropical Diseases.

REFERENCES

1. Butcher, G. A., Mitchell, G. H., and Cohen, S. (1973): Mechanism of host specificity in malarial infection. *Nature*, 244:40–41.
2. Byrne, G. I., and Moulder, J. W. (1978): Parasite-specified phagocytosis of *Chlamydia psittaci* and *Chlamydia trachomatis* by L and HeLa cells. *Infect. Immun.*, 19:598–606.
3. Chang, K. P., and Dwyer, D. M. (1976): Multiplication of a human parasite *(Leishmania donovani)* in phagolysosomes of hamster macrophages *in vitro*. *Science*, 193:678–680.
4. Goldman, R. (1977): Lectin-mediated attachment and ingestion of yeast cells and erythrocytes by hamster fibroblasts. *Exp. Cell. Res.*, 104:325–334.
5. Jensen, J. B., and Edgar, S. A. (1976): Effects of antiphagocytic agents on penetration of *Eimeria magna* sporozoites into cultured cells. *J. Parasitol.*, 62:203–206.
6. Jones, T. C., and Hirsch, J. G. (1972): The interaction between *Toxoplasma gondii* and mammalian cells. II. The absence of lysosomal fusion with phagocytic vacuoles containing living parasites. *J. Exp. Med.*, 136:1173–1194.
7. Jones, T. C., Yeh, S., and Hirsch, J. G. (1972): The interaction between *Toxoplasma gondii* and mammalian cells. I. Mechanism of entry and intracellular fate of the parasite. *J. Exp. Med.*, 136:1157–1172.
8. Lewis, J. T., Hafeman, D. G., and McConnell, H. M. (1980): Kinetics of antibody-dependent binding of haptenated phospholipid vesicles to a macrophage-related cell line. *Biochemistry*, 19:5376–5386.
9. Lonberg-Holm, K., and Whiteley, N. M. (1976): Physical and metabolic requirements for early interaction of poliovirus and human rhinovirus with HeLa cells. *J. Virol.*, 19:857–870.
10. Miller, L. H., Haynes, J. D., McAuliffe, F. M., Shiroishi, T., Durocher, J. R., and McGinniss, M. H. (1977): Evidence for differences in erythrocyte surface receptors for the malarial parasites, *Plasmodium falciparum* and *Plasmodium knowlesi*. *J. Exp. Med.*, 146:277–281.
11. Miller, L. H., Mason, S. J., Dvorak, J. A., McGinniss, M. H., and Rothman, I. K. (1975): Erythrocyte receptors for *(Plasmodium knowlesi)* malaria: Duffy blood group determinants. *Science*, 189:561–563.
12. Noble-Harvey, J., and Lonberg-Holm, K. (1974): Sequential steps in attachment of human rhinovirus type 2 to HeLa cells. *J. Gen. Virol.*, 25:83–91.
13. Schwartz, M. (1980): Interaction of phages with their receptor proteins. In: *Virus Receptors*, edited by L. L. Randall and L. Philipson, pp. 59–94. Chapman and Hall, London.
14. Volsky, D. J., Shapiro, I. M., and Klein, G. (1980): Transfer of Epstein-Barr virus receptors to receptor-negative cells. *Proc. Natl. Acad. Sci. USA*, 77:5453–5457.
15. White, J., Karlenbeck, J., and Helenius, A. (1980): Fusion of Semliki Forest virus with the plasma membrane can be induced by low pH. *J. Cell Biol.*, 87:264–272.
16. Zenian, A. (1981): *Leishmania tropica:* Biochemical aspects of promastigotes' attachment to macrophages *in vitro*. *Exp. Parasitol.*, 51:175–187.
17. Zenian, A., Rowles, P., and Gingell, D. (1979): Scanning electron-microscopic study of the uptake of *Leishmania* parasites by macrophages. *J. Cell Sci.*, 39:187–199.

Molecular Biology of Parasites, edited by
J. Guardiola, L. Luzzatto, and W. Trager.
Raven Press, New York © 1983.

In vitro Growth of Parasites

William Trager

The Rockefeller University, New York, New York 10021

The cultivation of parasitic eukaryotic organisms, with special reference to protozoa and helminths, has been reviewed briefly by me (85) and more extensively in three recent publications (48,70,82). I confine the discussion here to recent developments in the cultivation of parasitic protozoa, with emphasis on the uses to which cultures are being put. I do not consider cultivation of a parasitic organism as primarily an end in itself. Rather it is a means to many other ends.

It provides for approaches to problems otherwise difficult or impossible to study. In the simplest case, it permits the organism to be brought into laboratories where it otherwise could not be kept. For example, before the cultivation of *Plasmodium falciparum*, the only sources of this parasite were human infections, infections in splenectomized chimpanzees, or, since 1965, infections in certain small New World monkeys, particularly *Aotus trivirgatus*. Aotus monkeys are now in short supply and very expensive. But now that we can grow the parasite continuously in human erythrocytes *in vitro* (86), laboratories of molecular and cell biology that would not otherwise have studied *P. falciparum* are contributing to our understanding of this organism. Cultivation normally permits the production of larger amounts of parasite material than is otherwise obtainable. This is useful for some kinds of biochemical work and for immunological and serological studies. In perhaps its most important aspect, cultivation places the organism under more nearly controlled conditions. This control is greatly increased if the cultures are axenic and still more so if the organism can be grown in a defined medium. It is worth noting that in the groups *Apicomplexa* and the *Microspora* not one species has been grown axenically, that is, apart from its living host. And, as yet, relatively few species of parasitic protozoa have been grown in a defined medium, an essential condition for determining their nutritional requirements.

It is in relation to these uses of the cultures that I would like to discuss the work on cultivation of parasitic protozoa. I cannot think of a better group of organisms with which to begin than the trypanosomatid flagellates. Here, we find many different positive accomplishments and yet much scope for further work. The entire complex life cycles of trypanosomes have now been reproduced *in vitro*. It was less than five years ago that Hirumi (70) first obtained growth of the bloodstream forms of *Trypanosoma brucei in vitro* by using an appropriate culture medium combined with a suitable feeder layer of tissue culture cells. A modification of

Hirumi's method has been used to grow excellent cultures of *T. rhodesiense* bloodstream forms (36), but no continuous cultures at 37°C of the bloodstream form of *T. congolense* have yet been reported, though such forms have been maintained up to 21 days in dermal explants (34). By manipulating temperature and feeding of the cultures of *T. brucei*, it was possible to produce procyclic forms, epimastigotes, and metacyclic forms at 25°C and again get bloodstream trypanomastigotes when the temperature was raised to 37°C. The precise conditions determining this cyclic development (70) have not been identified. Lowering the temperature to 25°C was essential to initiate it. The cultures were held without change of medium and without opening the flasks; many trypanosomes died, so that their products of disintegration might have had an influence. In any case, midgut and proventricular forms appeared and the cultures became noninfective to mice. When medium changes were again begun with the cultures still at 25°C, small numbers of epimastigotes and finally of metacyclic forms appeared, and the cultures were again infective to mice. After 40 days of such low-temperature growth, 26 out of 30 cultures had reacquired infectivity. If such cultures were again incubated at 37°C, typical infective bloodstream trypomastigotes reappeared and multiplied rapidly.

These results would indicate that tsetse fly organs may not be essential for development of the insect cycle. However, tsetse fly tissues, in particular, head and salivary glands, will support *in vitro* more complete cyclic development from procyclic culture forms to epimastigotes to metacyclic trypanosomes that are infective to mice (17). Cultures prepared with 25 or fewer head-salivary gland explants rarely became infective, but those with 27 to 50 such explants consistently produced infective trypanosomes. Antigenic analysis of these metacyclic trypanosomes by immunofluorescence shows that they correspond to those produced in tsetse flies (32). Similar results have now been obtained with *T. congolense* (35) by initiating cultures with trypanosomes from the mouthparts of infected *Glossina mortisans* placed with bovine dermal collagen explant at 28°C. Primary cultures became infective for mice after two weeks and remained infective continuously up to 76 days. Subcultures made into medium with proboscides from uninfected flies remained infective up to 215 days. The cultures produced as many as 1.6×10^6 infective organisms every 2 days, and all the morphological forms normally seen in the proboscis of infected flies were present. It is clear that we may be approaching the time when it will be possible to define and manipulate the conditions controlling the differentiations that characterize the complex life cycles of the African trypanosomes.

Cultures of the bloodstream forms of *T. brucei* are meanwhile already in use for study of the remarkable phenomenon of antigenic variation. This variation occurs *in vitro* much as it does *in vivo*, thus finally ruling out any role of the immune system of the host in the induction of antigenic variation (25). Tissue culture systems very like that of Hirumi for *T. brucei* have also been used in the cultivation of bloodstream forms at 37°C of the rather host-specific trypanosomes of rodents, *T. lewisi*, *T. acomys*, and *T. musculi* (1,27,90).

Among the stercorarian trypanosomes of mammals, *T. theileri* has repeatedly been observed to grow in tissue culture at 37°C with formation of trypomastigotes (50). In general with such trypanosomes, cultures grown at 26°C reproduce the invertebrate cycle, including the formation of infective forms. This is true also of trypanosomes of birds [as *Trypanosoma avium* (3)] and of trypanosomes of fish and other cold-blooded vertebrates (67). With the latter, the interesting problem remains of producing *in vitro* the large striated blood forms. With many of the stercorarian trypanosomes, bloodstream forms have been obtained only in the presence of a feeder layer of tissue culture cells, as is true with the salivarian forms. At least one strain of *T. theileri*, however, has been grown continuously at 37°C in a semidefined medium (77).

T. cruzi, the medically most important of the stercorarian trypanosomes, will readily produce its vertebrate stages, intracellular amastigotes maturing to trypomastigotes, in a variety of tissue cultures. This has been known since the early work of Meyer (54) and Neva (59). More recently, however, all the stages of the life cycle of *T. cruzi* have been obtained in nonliving media, that is, axenically, at 37°C (64,65).

Whereas the amastigotes of *T. cruzi*, its intracellular multiplicative form, will develop in nonliving media, the superficially similar amastigotes of *Leishmania* have never yet been grown extracellularly. It is interesting to note that despite superficial similarities of the amastigotes of *T. cruzi* and *Leishmania*, their physiology must be very different. Amastigotes of *T. cruzi* develop directly in the cytoplasm of host cells such as muscle or fibroblasts, whereas those of *Leishmania* develop in the lysosomal vacuole of macrophages. While many investigators have obtained brief development of various species of *Leishmania* within macrophages *in vitro*, only Lamy and others using his dog histiocytoma line have obtained continuous cultivation of *L. donovani* amastigotes (45). More recently, Chang et al. (11) have developed a still better system for continuous *in vitro* production of amastigotes of *L. mexicana amazonensis* using the mouse macrophage line J744EB.

In sharp contrast to the difficulty of *in vitro* growth of amastigotes of *Leishmania* is the apparent ease with which the promastigotes of many species can be grown. Here, recent progress is concerned with growth in defined media, and this will be considered later. There is another aspect, however, of the growth of promastigotes of *Leishmania* to which too little attention has been given, namely, their infectivity. Many cultures of promastigotes are said to lose infectivity completely, whereas others are more or less infective. There are indications that the previous conditions of culture affect infectivity (33), but this is only now being investigated in more detail (J. Keithly, personal communication). It is possible that promastigotes in sandflies may undergo developmental changes culminating in high infectivity and that we do not yet know how to produce such forms *in vitro*. Several insect cell culture media have proved useful for culture of leishmanial promastigotes (70).

The fact that some differentiated stages of the life cycle do not occur under ordinary *in vitro* conditions points to the value of cultures in studies on the developmental cycle. By manipulating the culture conditions, one can try to find out what conditions are responsible for the differentiated development. A simple illustration is provided by the encystation of *Entameba invadens*. Axenic cultures of this amoeba can be made to encyst by placing them in an appropriate medium (69). On the other hand, encystation (14,23) of axenically grown *E. histolytica* has never been obtained, despite many efforts (22,23). The formation of round cystlike bodies with a thick outer wall was observed when axenic *E. histolytica* were fed starch, exposed to bacterial endotoxin, and then placed in a minimal medium with epinephrine and theophylline at 33°C (56). When placed in fresh growth medium, these bodies hatched to yield trophic forms. Of the several species of the intestinal flagellate *Giardia* that have been grown axenically (52,53), none has been seen to encyst in such cultures. Excystation, however, occurs readily if the cysts are first exposed to a synthetic gastric juice of pH 1.3 to 2.7 and then placed in axenic culture medium (5).

Parasites in culture may lose infectivity or pathogenicity for the host. If infectivity depends on differentiation of special infective forms, as the metacyclic forms of trypanosomes, and if these forms are not produced in the cultures, the lack of infectivity is readily explained. But in other instances, no ready explanation is available. For example, *E. histolytica* when grown axenically loses its pathogenicity for experimental animals (21), and this is only slowly restored when the amoebae are grown with bacterial associates. While this seemed rather mysterious, some of the mystery has been removed by the recent finding that pathogenicity could be restored to cultures grown axenically for 5 to 6 years by supplementing the medium with cholesterol (51). The number of passages needed in cholesterol-supplemented medium to restore a certain degree of pathogenicity for hamsters was proportional to total period in culture, being much larger for a line isolated in 1924 than for one isolated in 1967. Once restored, pathogenicity persisted for a long time after cholesterol treatment was stopped.

Of special interest with regard to loss of infectivity and pathogenicity in culture have been the studies of Honigberg with several species of trichomonads (39). A strain of *Trichomonas gallinae* highly pathogenic for pigeons lost pathogenicity after 17 to 21 weeks of axenic culture, and some strains even lost their infectivity. If the rate of multiplication in culture was slowed by omitting carbohydrate from the medium, the rate of loss of pathogenicity was also slowed, suggesting the gradual dilution of a cytoplasmic factor. In keeping with this was the finding that an attentuated strain exposed to both DNA and RNA from a virulent strain regained some pathogenicity, as measured by a mouse assay, and also regained infectivity for pigeons. However, when cultures of *T. gallinae* were grown for over a year in the presence of chick liver cell cultures, there was no loss in virulence, despite a high growth rate. Evidently, under these conditions the presumed cytoplasmic factors are not diluted away. It would be interesting to know precisely the factors derived from the tissue culture responsible for this effect. The results with cholesterol and *Entamoeba* suggest some approaches to this problem. Results similar to those with *T. gallinae* have been obtained with *T. vaginalis*, the human parasite. Prolonged axenic cultivation results in the loss of pathogenicity for mice.

The ultimate goal of culture work is to develop a defined medium, or a series of defined media, in which the parasite will develop axenically through its entire life cycle and will retain infectivity. This goal has been achieved for some of the trypanosomatid flagellates that inhabit the alimentary tract of insects. Several defined media have been developed, but it is of interest and very useful that the single medium of Steiger and Steiger (79), devised for the promastigotes of *Leishmania donovani*, was found to support excellent growth of three species of *Leptomonas*, four of *Herpetemonas*, six of *Crithidia*, and one of *Blastocrithidia*, all from insects (30). This same medium will also support the growth of the plant flagellate *Phytomonas davidi* [only recently placed in culture (49)], as well as of the promastigotes of *Leishmania tarentolae*, *L. braziliensis*, and *L. donovani*. This medium therefore deserves to be examined (Table 1). Cystine could be omitted as long as cysteine was present. Glutamic acid could also be omitted (80), as well as serine. It is clear that we have here a medium for the study of comparative nutritional requirements of a range of lower trypanosomatids and of promastigotes of several species of *Leishmania*, including two important pathogens of man. Such a medium and the similar medium of Berens and Marr (4), prepared in part from commercially available solutions, lend themselves to detailed study of nutritional requirements, and some work of this sort has already been done with *L. donovani* promastigotes

TABLE 1. *Composition of medium RE I (mg/liter)*

A.	NaCl	8000	C.	$NaHCO_2$	1000
	KCl	400		HEPES	14,250
	$MgSO_4 \cdot 7H_2O$	200	D.	Adenosine	20
	$Na_2HPO_4 \cdot 2H_2O$	60		Guanosine	20
	KH_2PO_4	60	E.	D-Biotin	1
	$CaCl_2$	70		Choline cloride	1
	Glucose	2000		Folic acid	11
	Sodium Acetate	600		*i*-Inositol	2
B.	L-Arginine HCl	200		Niacinamide	1
	L-Cysteine HCl	50		D-Calcium pantothenate	1
	L-Cystine	50		Pyridoxal HCl	1
	L-Glutamic acid	300		Riboflavin	0.1
	L-Glutamine	300		Thiamine HCl	1
	L-Histidine	100	F.	Lipoic acid	0.4
	L-Isoleucine	100		Menadione	0.4
	L-Leucine	300		Vitamin A	0.4
	L-Lysine HCl	250	G.	Ascorbic acid	0.2
	L-Methionine	50		Vitamin B_{12}	0.2
	L-Phenylalanine	100		Bovine albumin	15
	L-Proline	300		(defatted)	
	DL-Serine	200		Hemin	10
	L-Threonine	400		Phenol red	10
	L-Tryptophan	50		Redistilled H_2O Q.S.	1000 ml
	L-Tyrosine	50		(pH adjusted with	
	L-Valine	100		1*N* NaOH to 7.3–7.4)	

(From ref. 79, with permission.)

(78,80). It is noteworthy that either proline or glucose could be omitted but not both, parallelling results of Krassner and Flory with *L. tarentolae* and again emphasizing the importance of proline in the metabolism of hemoflagellates (7,84). *T. cruzi* has recently been grown in a defined medium at 27°C using purified bovine liver catalase as source for both hemin and amino acids (2). Five different strains not only grew but also produced small numbers (5%) of infective trypomastigote forms. Other hemin-containing proteins could not be substituted for bovine liver catalase. Subunits of bovine liver catalase prepared by treatment with 6*M* urea still supported growth, but an acid hydrolysate of the protein, or a mixture of its constituent amino acids, did not. Procyclic forms of certain strains of *T. brucei* have been grown in a nearly defined medium (16) containing casein hydrolysate.

Among the insect hemoflagellates that have been grown in defined media are three species that harbor intracellular bacteria: *Crithidia oncopelti*, *C. deanei*, and *Blastocrithidia culicis*. When *C. oncopelti* and *B. culicis* were rendered aposymbiotic, they could no longer be grown in defined medium; they now showed a requirement for a factor present in liver and as yet unidentified (9,11). *C. deanei* freed of its symbiotes showed a similar requirement, but this could be replaced with a high level of nicotinamide, 3 to 5 mg/100 ml as compared with the usual level in defined media of 1 to 5 mg/liter (57). Numerous other factors, including hemin, are supplied by the symbiotic bacteria, so that symbiote-containing strains can be grown in relatively very simple media (58). The symbiotic bacteria have been shown to supply enzymes for heme synthesis in *C. oncopelti* and *B. culicis* (11) and of the urea cycle in *B. culicis* and *C. deanei* (31).

The only group of parasitic protozoa other than trypanosomatids for which even nearly defined media have been available are the trichomonads (38,74). These flagellates appear to have more complex nutritional requirements than the hemoflagellates. These include cholesterol, fatty acids, and fat-soluble vitamins, as well as the usual array of water-soluble vitamins and amino acids. Attempts to correlate nutritional factors with loss or recovery of pathogenicity would be of special interest for this group.

It is only recently that representative species of the symbiotic cellulose-digesting flagellates of termites and of the wood-feeding roach *Cryptocercus punctulatus* have been grown axenically (92,94), making possible detailed studies of their cellulose metabolism (93). *Trichomitopsis* was first obtained in axenic culture by the use of strict anaerobiosis and a medium supplemented with autoclaved rumen bacteria. The same methods were then successfully applied to other species, including the hypermastigotes *Trichonympha spherica* and a *Trichonympha* from *Cryptocercus* (M. Yamin, personal communication). It seems possible that similar methods might permit axenic cultures of the rumen ciliates. Here, many species of entodiniomorphids have been cultured with a mixed bacterial flora (15,55), and *Entodinium caudatum* has been maintained in monoxenic culture for over two months (37). The holotrich rumen protozoa have proved difficult to culture, but they have been kept going as mixed cultures in rumen fluid (18).

As I mentioned at the beginning, there are two large groups of entirely parasitic protozoa, none of which has been grown axenically. These are the *Apicomplexa* and the *Microspora*. Many species spend all or part of their developmental cycle as parasites within other living cells. With the development of tissue culture methods, a number of species have been propagated in cell cultures. Several kinds of *Microspora* have been grown in insect cell cultures, as well as in vertebrate cell cultures (6,89). Whereas some species develop through only one cycle, from spore back to spore (6), others show successive cycles. The latter is seen especially with *Nosema (Encephalitozoon) cuniculi*, a parasite of vertebrates (63). With this species, intracellular extrusion of the polar filament and sporoplasm were observed, and this may account for the continuous propagation. Other propagative forms may, however, be present—as in the hemolymph of silkworms infected with *N. bombycis* (62).

The tachyzoite stage of *Toxoplasma gondii* (26) has long been grown in a number of different kinds of cell cultures, and this stage can be propagated indefinitely in this way. Such cultures have been used to study the effects of drugs (73), parasite-host cell interactions (47), and immunity (43), and parasite metabolic pathways (66). With other species of coccidia that lack a continuously propagating asexual form, as in *Eimeria*, cell cultures support development of first-generation merozoites from sporozoites and sometimes of second-generation merozoites (71,83). For two species of *Eineria* development through to oocysts has been obtained *in vitro*, first in *E. tenella* (24) and then in *E. bovis* (76). It is highly interesting that gametogony of *Sarcocystis* with proof of its coccidial nature was first observed *in vitro* when Fayer (29) inoculated bradyzoites (26) from *Sarcocystis* cysts in a grackle into cell culture. All of this serves to emphasize again the special value of cultures in relation to developmental cycles.

With several kinds of parasites of the class Sporozoea in the *Haemosporina* and the *Piroplasmida* (46), portions of the life cycle have been grown in cell cultures *in vitro*. For the piroplasmids *Theileria parva*, the cause of East Coast fever of cattle, and *T. annulata*, cause of a less severe but still important disease, the macroschizonts have been cultured. These develop within lymphoblastoid cells transformed by the parasites and capable of continuous growth (8). Division of the host cells depends on the presence of the parasite. At cell division the parasites are distributed evenly to each daughter cell. Lymphoid cells from cattle can be infected *in vitro* with sporozoites from infected ticks; once infected, they are transformed and can then be propagated together with the macroschizonts within them. Such cultures have been used for a successful vaccination of cattle against *T. annulata* and as source of antigens for serodiagnostic and immunologic work. Microschizonts, the forms that give rise to the stage that infects the erythrocytes, have been produced in the cultures, and infection of a small proportion of added red cells has been seen *in vitro* (19,20,41). The piroplasms in erythrocytes are the presumed gametocytes that initiate a cycle in the vector tick, but *in vitro* development of this cycle has not yet been achieved.

With the other important group of piroplasmids, the *Babesias*, the situation appears superficially simpler. The erythrocytic stages of *B. bovis*, the most important babesia of cattle, have been grown continuously in cattle red cells (28), best results having been obtained with a static system under low O_2 tension (42), much like that used for *P. falciparum*. Material from the supernatants of such cultures has been used to immunize cattle (75). There is no evidence that the cultures are infective to ticks.

We come now to consideration of malaria parasites, from the standpoint of human disease, the most important members of this group of protozoa. It is just five years since the first continuous culture of any malaria parasite. This was the human parasite, *P. falciparum* (87). Since then, the erythrocytic stages of four other species of malaria parasites, all from rhesus monkeys, have been cultured by the same methods. These are *P. knowlesi* (12,91) with a 24-hr cycle, and a vivax-like parasite, *P. cynomolgi* (61). In addition, there has been a report of the parasites of human quartan malaria *P. malariae* in mixed culture with *P. falciparum* (68). In view of the successful culture of *P. cynomolgi* by the same methods as for *P. falciparum*, it is curious that there has been no success with *P. vivax* as yet.

For obvious reasons, most work continues to be done with *P. falciparum*, and this parasite is now maintained in culture in a number of laboratories throughout the world. The cultures are being used to screen for antimalarials, to study drug resistance, and to investigate host-parasite relations. They are providing information about the nature of the receptors on the host erythrocyte responsible for attachment by the merozoite, and why certain genetic red cell variants, as HbS, provide relative resistance to falciparum malaria. Most importantly, the cultures are being used in studies that we hope will lead to the identification of antigens responsible for protective immunity and in this way to a vaccine against malaria (86).

We must remember that we have *in vitro* only one of the three cycles of development that comprise the complete life cycle of malaria parasites. This is the erythrocytic cycle. In this cycle in nature certain individual merozoites, for unknown reasons, develop into gametocytes rather than continuing the asexual cycle. These male and female gametocytes cannot develop further in the vertebrate host, but if ingested by a suitable mosquito, they produce gametes that initiate the sexual or sporogonic cycle. Formation of gametocytes was noted in the early cultures of falciparum. It was found that conditions deleterious to the axesual forms favored appearance of gametocytes, and there is evidence for the involvement of cyclic adenosine monophosphate in this differentiation (44). In all early work, the gametocytes were noninfective to mosquitoes. Recently, however, gametocytes infective to mosquitoes have been regularly produced when hypoxanthine was added to the medium (T. Ifediba and J. Vanderberg, *personal communication*). Thus, the cultures can now provide a source of infected mosquitoes. Ookinete formation and partial development of already formed oocytes have been obtained *in vitro*, but nothing approaching the full development from ookinete through oocyst to infective sporozoites (72).

The sporozoites initiate the third developmental cycle of malaria. In bird malaria, the preerythrocytic or exoerythrocytic cycle occurs in endothelial cells. It has been propagated continuously in tissue cultures since the pioneer work of Huff (40,88). In mammalian malaria, however, the preerythrocytic cycle occurs in liver hepatic cells and is self-limited, that is, it does not repeat but must give rise to the erythrocytic cycle. Recently, this cycle has been obtained *in vitro* for the rodent malaria agent, *P. berghei* (80), using sporozoites placed in appropriate tissue cultures. It is interesting that *in vitro* there was no requirement for hepatic cells; the best cultures were in human embryonic lung cells. No similar development of a preerythrocytic stage of a human malaria parasite has yet been obtained, but there is now reason to hope that this can be done.

With these intracellular protozoa, as with other obligate intracellular parasites, there remains the most challenging problem of all—to grow them axenically, that is, without the host cell in a nonliving medium. The erythrocytic stages of malaria may be supposed to be exceptionally favorable material for such work, since they develop in a host cell that is no longer synthesizing protein. And indeed some extracellular development of the bird malaria agent, *P. lophurae*, has been obtained (88). Similar experiments with *P. falciparum* might be even more productive, since more is known about the human erythrocyte than any other type of cell.

REFERENCES

1. Albright, J. W., and Albright, J. F. (1978): Growth of *Trypanosoma musculi* in cultures of murine spleen cells and analysis of the requirement for supportive spleen cells. *Infect. Immun.*, 22:343–349.
2. Avila, J. L., Bretana, A., Casanova, M. A., Avila, A., and Rodriguez, F. (1979): *Trypanosoma cruzi:* Defined medium for continuous cultivation of virulent parasites. *Exp. Parasitol.*, 48:27–35.
3. Baker, J. R. (1966): Studies on *Trypanosoma avium*. IV. The development of infective metacyclic trypanosomes in cultures grown *in vitro*. *Parasitology*, 56:15–19.
4. Berens, R. L., and Marr, J. J. 1978): An easily prepared defined medium for cultivation of *Leishmania donovani* promastigotes. *J. Parasitol.*, 64:160.
5. Bingham, A. K., and Meyer, E. A. (1979): Giardia excystation can be induced *in vitro* in acidic solutions. *Nature*, 277:301–302.
6. Bismanis, J. E. (1970): Detection of latent murine nosematosis and growth of *Nosema cuniculi* in cell cultures. *Can. J. Microbiol.*, 16:237–242.
7. Bowman, I. B. R. (1974): Intermediary metabolism of pathogenic flagellates. In: *Ciba Foundation Symposium 20. Trypanosomiasis and Leishmaniasis*. pp. 255–270. Elsevier/North-Holland.
8. Brown, C. G. D. (1980): *In vitro* cultivation of Theileria. In: *Tropical Disease Research Series No. 3: The* in vitro *Cultivation of the Pathogens of Tropical Diseases*, pp. 127–144. Schwabe and Co., Basel.
9. Chang, K.-P. (1976): Symbiote-free hemoflagellates, *Blastocrithidia culicis* and *Crithidia oncopelti:* Their liver factor requirement and serologic identity. *J. Protozool.*, 23:241–244.
10. Chang, K.-P. (1980): Human cutaneous leishmania in a macrophage line: Propagation and isolation of intracellular parasites. *Science*, 209:1240–1242.
11. Chang, K.-P., Chang, C. S., and Sassa, S. (1975): Heme biosynthesis in bacterium-protozoon symbiosis: Enzyme defects in host hemoflagellates and complemental role of their intracellular symbiotes. *Proc. Natl. Acad. Sci. USA*, 72:2979–2983.
12. Chen Zhengren, Gao Minxin, Li Yuhua, Han Shumin, and Zhang Nailin (1980): Studies on the cultivation of erythrocytic stage Plasmodium *in vitro*. *Chin. Med. J.*, 93:31–35.
13. Chin, W., Moss, De Z., and Collins, W. E. (1979): The continuous cultivation of *Plasmodium fragile* by the method of Trager-Jensen. *Am. J. Trop. Med. Hyg.*, 28:591–592.

14. Cleveland, L. R., and Sanders, E. P. (1930): Encystation, multiple fission without encystment, excystation, metacyclic development, and variation in a pure line and strains of *Entamoeba histolytica*. *Arch. Protistenk.*, 70:223–266.
15. Coleman, G. S. (1979): Rumen ciliate protozoa. In: *Biochemistry and Physiology of Protozoa*, 2nd ed. Vol. 2, edited by M. Levandowsky and S. H. Hutner, pp. 381–408. Academic Press, New York.
16. Cross, G. A. M., and Manning, J. C. (1973): Cultivation of *Trypanosoma brucei* sspp. in semi-defined and defined media. *Parasitology*, 67:315–331.
17. Cunningham, I., and Taylor, A. M. (1979): Infectivity of *Trypanosoma brucei* cultivated at 28° with tsetse fly salivary glands. *J. Protozool.*, 26:428–432.
18. Czerkawski, J. W., and Breckenridge, G. (1977): Design and development of a long-term rumen simulation technique (Rusitec). *Br. J. Nutr.*, 38:371–384.
19. Danskin, D., and Wilde, J. K. H. (1976): Simulation *in vitro* of bovine host cycle of *Theileria parva*. *Nature*, 261:311–312.
20. Danskin, D., and Wilde, J. K. H. (1976): The effect of calf lymph and bovine red blood cells on *in vitro* cultivation of *Theileria parva*-infected lymphoid cells. *Trop. Anim. Health Prod.*, 8:175–185.
21. Das, S. R., Das, P., and Rai, G. P. (1979): Revival of pathogenicity of axenically grown *Entamoeba histolytica* for the rat. *Aust. J. Exp. Biol. Med. Sci.*, 57:241–244.
22. Diamond, L. S. (1961): Axenic cultivation of *Entamoeba histolytica*. *Science*, 134:336–337.
23. Diamond, L. S. (1980): Axenic cultivation of *Entamoeba histolytica:* progress and problems. *Arch. Invest. Med. (Mex.) (Suppl. 1)*, 11:47–54.
24. Doran, D. J. (1971): Increasing the yield of *Eimeria tenella* oocysts in cell cultures. *J. Parastiol.*, 57:891–900.
25. Doyle, J. J., Hirumi, H., Hirumi, K., Lupton, E. N., and Cross, G. A. M. (1980): Antigenic variation in clones of animal-infective *Trypanosoma brucei* derived and maintained *in vitro*. *Parasitology*, 80:359–369.
26. Dubey, J. P. (1977): *Toxoplasma*, *Hammondia*, *Besnoitia*, *Sarcocystis* and other tissue cyst-forming coccidia of man and animals. In: *Parasitic Protozoa*, Vol. 3, edited by J. P. Kreier, pp. 107–237. Academic Press, New York.
27. El On, J., and Greenblatt, C. L. (1977): *Trypanosoma lewisi*, *T. acomys* and *T. cruzi:* A method for their cultivation with mammalian tissue. *Exp. Parasitol.*, 41:31–42.
28. Erp, E. E., Smith, R. D., Ristic, M., and Osomo, B. M. (1980): Continuous *in vitro* cultivation of *Babesia bovis*. *Am. J. Vet. Res.*, 41:1141–1142.
29. Fayer, R. (1972): Gametogony of *Sarcocystis* sp. in cell culture. *Science*, 175:65–67.
30. Fish, W. R., Holz, G. G., Jr., and Beach, D. H. 1978): Cultivation of trypanosomatids. *J. Parasitol.*, 64:546–547.
31. Galinari, S., and Camargo, E. P. (1979): Urea cycle enzymes in wild and aposymbiotic strains of *Blastocrithidia culicis*. *J. Parasitol.*, 65:88.
32. Gardiner, P. R., Jones, T. W., and Cunningham, I. (1980): Antigenic analysis by immunofluorescence of *in vitro*-produced metacyclics of *Trypanosoma brucei* and their infections in mice. *J. Protozool.*, 27:316–320.
33. Giannini, M. S. (1974): Effects of promastigote growth phase, frequency of subculture and host age on promastigote-initiated infections with *Leishmania donovani* in the golden hamster. *J. Protozool.*, 21:521–527.
34. Gray, M. A., Brown, C. G. D., Luckins, A. G., and Gray, A. R. (1979): Maintenance of infectivity of *Trypanosoma congolense in vitro* with explants of infected skin at 37°. *Tran. R. Soc. Trop. Med. Hyg.*, 73:406–408.
35. Gray, M. A., Cunningham, I., Gardiner, P. R., Taylor, A. M., and Luckins, A. G. (1981): Cultivation of infective forms of *Trypanosoma congolense* from trypanosomes in the proboscis of *Glossina morsitans*. *Parasitology*, 82:81–95.
36. Hill, G. C., Shimer, S. P., Caughey, B., and Sauer, L. S. (1978): Growth of infective forms of *Trypanosoma rhodesiense in vitro*, the causative agent of African trypanosomiasis. *Science*, 202:763–765.
37. Hino, T., and Kanetaka, M. (1977): Gnotobiotic and axenic cultures of a rumen protozoan, *Entodinium caudatum*. *J. Gen. Appl. Microbiol.*, 23:37–48.
38. Honigberg, B. M. (1978): Trichomonads of veterinary importance. Trichomonads of importance in human medicine. In: *Parasitic Protozoa*, Vol. 2, edited by J. Kreier, pp. 163–273, 275–454. Academic Press, New York.

39. Honigberg, B. M. (1979): Biological and physiological factors affecting pathogenicity of trichomonads. In: *Biochemistry and Physiology of Protozoa*, 2nd ed., Vol. 2, edited by M. Lewandowsky and S. H. Hutner, pp. 409–427. Academic Press, New York.
40. Huff, C. G. (1964): Cultivation of the exoerythrocytic stages of malarial parasites. *Am. J. Trop. Med. Hyg.*, 13:171–177.
41. Hulliger, L., Brown, C. G. D., and Wilde, J. K. H. (1966): Transition of developmental stages of *Theileria parva in vitro* at high temperature. *Nature*, 211:328–329.
42. James, M. A., Levy, M. G., and Ristic, M. (1981): Isolation and partial characterization of culture-derived soluble *Babesia bovis* antigens. *Infect. Immun.*, 31:358–361.
43. Jones, T. C., Len, L., and Hirsch, J. G. (1975): Assessment *in vitro* of immunity against *Toxoplasma gondii*. *J. Exp. Med.*, 141:466–482.
44. Kaushal, D. C., Carter, R., and Miller, L. H. (1980): Gametocytogenesis by malaria parasites in continuous culture. *Nature*, 286:490–492.
45. Lamy, L. H. (1972): Protozoaires intracellulaires en culture cellulaire: Interet-possibilities-limites. *Annee Biol.*, 11:145–183.
46. Levine, N. D., *et al.* (1980): A newly revised classification of the Protozoa. *J. Protozool.*, 27:37–58.
47. Lycke, E., Carlberg, K., and Norrby, R. (1976): Interactions between *Toxoplasma gondii* and its host cells: function of the penetration-enhancing factor of Toxoplasma. *Infect. Immun.*, 11:853–861.
48. Maramorosch, K., and Hirumi, H., eds. (1979): *Practical Tissue Culture Applications*. Academic Press, New York.
49. McGhee, R. B., and Postell, F. J. (1976): Axenic cultivation of *Phytomonas davidi* (Trypanosomatidae), a symbiote of laticiferous plants. *J. Protozool.*, 23:238–241.
50. McHolland-Raymond, L. E., Kingston, N., and Trublood, M. S. (1978): Continuous cultivation of *Trypanosoma theileri*, at 37°C in bovine cell culture. *J. Protozool.*, 25:388–394.
51. Meerovitch, E., and Ghadirian, E. (1978): Restoration of virulence of axenically cultivated *Entamoeba histolytica* by cholesterol. *Can. J. Microbiol.*, 24:63–65.
52. Meyer, E. A. (1970): Isolation and axenic cultivation of *Giardia* trophozoites from the rabbit, chinchilla, and cat. *Exp. Parastiol.*, 27:179–183.
53. Meyer, E. A. (1976): *Giardia lamblia:* isolation and axenic cultivation. *Exp. Parasitol.*, 39:101–105.
54. Meyer, H., and Xavier de Oliveira, M. (1948): Cultivation of *Trypanosoma cruzi* in tissue cultures: a four year study. *Parasitology*, 39:91–94.
55. Michelowski, T. (1979): A simple system for continuous culture of rumen ciliates. *Bull. Acad. Polon. Sci., Sci. Biol.*, 27:581–583.
56. Mitra, S., and Murti, C. B. K. (1978): Encystation of axenically grown *Entamoeba histolytica:* effect of bacterial endotoxins, starch and epinephrine. *Proc. Ind. Acad. Sci. B.*, 87:9–23.
57. Mundim, M. H., and Roitman, I. (1977): Extra nutritional requirements of artificially aposymbiotic *Crithidia deanei*. *J. Protozool.*, 24:329–331.
58. Mundim, M. H., Roitman, I., Herman, M. A., and Kitajima, E. W. (1974): Simple nutrition of *Crithidia deanei*, a reduviid trypanosomatid with an endosymbiont. *J. Protozool.*, 21:518–521.
59. Neva, F. A., Malone, M. F., and Myers, B. R. (1961): Factors influencing the intracellular growth of *Trypanosoma cruzi in vitro*. *Am. J. Trop. Med. Hyg.*, 10:140–154.
60. Nguyen-Dinh, P., Campbell, C. C., and Collins, W. E. (1980): Cultivation *in vitro* of the quartan malaria parasite *Plasmodium inui*. *Science*, 209:1249–1251.
61. Nguyen-Dinh, P., Gardner, A. L., Campbell, C. C., Skinner, J. C., and Collins, W. E. (1981): Cultivation *in vitro* of the vivax-type malaria parasite *Plasmodium cynomolgi*. *Science*, 212:1146–1148.
62. Ohshima, K. (1975): Propagative form of *Nosema bombycis* found in the hemolymph of silkworms. *Annot. Zool. J.*, 48:21–33.
63. Pakes, S. P., Shadduck, J. A., and Cali, A. (1975): Fine structure of *Encephalitozoon cuniculi* from rabbits, mice and hamsters. *J. Protozool.*, 22:481–488.
64. Pan, C. T. (1971): Cultivation and morphogenesis of *Trypanosoma cruzi* in improved liquid media. *J. Protozool.*, 18:556–560.
65. Pan, S. C. T. (1978): *Trypanosoma cruzi:* Intracellular stages grown in a cell-free medium at 37°C. *Exp. Parasitol.*, 45:215–224.
66. Pfefferkorn, E. R. (1978): The enzymic defect of a mutant resistant to 5-fluorodeoxyuridine. *Exp. Parasitol.*, 44:26–35.

67. Qadri, S. S. (1962): The development in culture of *Trypanosoma striati* from an Indian fish. *Parasitology*, 52:229–235.
68. Rai Chowdhury, A. N., Chowdhury, D. S., and Regis, M. L. (1979): Simultaneous propagation of *P. malariae* and *P. falciparum* in a continuous culture. *Ind. J. Med. Res. (Suppl.)*, 70:72–78.
69. Rengpian, S., and Bailey, G. B. (1975): Differentiation of Entamoeba: A new medium and optimal conditions for axenic encystation of *E. invadens*. *J. Parasitol.*, 61:24–30.
70. Rowe, D. S., and Hirumi, H., eds. (1980): *Tropical Diseases Research Series 3: The* in vitro *Cultivation of the Pathogens of Tropical Diseases*. UNDP/World Bank/WHO Special Programme for Research and Training in Tropical Diseases). Schwabe and Co., Basel.
71. Ruff, M. D., and Reid, W. M. (1977): Avian coccidia. In: *Parasitic Protozoa*, Vol. 3, edited by J. P. Kreier, pp. 34–69. Academic Press, New York.
72. Schneider, I., and Vanderberg, J. P. (1980): Culture of the invertebrate stages of plasmodium and the culture of mosquito tissues. In: *Malaria*, Vol. 2, edited by J. P. Kreier, pp. 235–270. Academic Press, New York.
73. Sheffield, H. G., and Melton, M. T. (1975): Effect of pyrimethamine and sulfadiazine on the fine structure and multiplication of *Toxoplasma gondii*. *J. Parasitol.*, 61:704–712.
74. Shorb, M. S. (1964): The physiology of trichomonads. In: *Biochemistry and Physiology of Protozoa*, Vol. 3, edited by S. H. Hutner, pp. 383–457. Academic Press, New York.
75. Smith, R. D., James, M. A., Ristic, M., Aikawa, M., and Vege y Murguia, C. A. (1981): Bovine babesiosos: Protection of cattle with culture-derived soluble *Babesia bovis* antigens. *Science*, 212:335–338.
76. Speer, C. A., and Hammond, D. M. (1973): Development of second generation schizonts, gametes and oocysts of *Eimeria bovis* in bovine kidney cells. *Z. Parasitenk.*, 42:105–113.
77. Sollard, A. E., and Soulsby, E. J. L. (1968): Cultivation of *Trypanosoma theileri* at 37° C in partially defined media. *J. Protozool.*, 15:463–466.
78. Steiger, R. F., and Black, C. D. V. (1980): Simplified defined media for cultivating *Leishmania donovani* promastigotes. *Acta Trop.*, 37:195–198.
79. Steiger, R. F., and Steiger, E. (1976): A defined medium for cultivating *Leishmania donovani* and *L. braziliensis*. *J. Parasitol.*, 62:1010–1011.
80. Steiger, R. F., and Steiger, E. (1977): Cultivation of *Leishmania donovani* and *L. braziliensis* in defined media: Nutritional requirements. *J. Protozool.*, 24:437–441.
81. Strome, C. P. A., De Santis, P. L., and Beaudoin, R. L. (1979): Cultivation of the exoerythrocytic stages of *Plasmodium berghei* from sporozoites. *In Vitro*, 15:531–536.
82. Taylor, A. E. R., and Baker, J. R., eds. (1978): *Methods of Cultivating Parasites* in vitro. Academic Press, London.
83. Todd, K. S., Jr., and Ernst, J. V. (1977): Coccidia of mammals except man. In: *Parasitic Protozoa*, Vol. 3, edited by J. P. Kreier, pp. 71–99. Academic Press New York.
84. Trager, W. (1974): Nutrition and biosynthetic capabilities of flagellates: Problems of *in vitro* cultivation and differentiation. *Ciba Foundation Symposium 20: Trypanosomiasis and Leishmaniasis*, pp. 225–245. Elsevier/North-Holland, Amsterdam.
85. Trager, W. (1978): Cultivation of parasites *in vitro*. *Am. J. Trop. Med. Hyg.*, 27:216–222.
86. Trager, W. (1980): Cultivation of malaria parasites *in vitro*: Its application to chemotherapy, immunology and the study of parasite-host interactions. In: *The Host Invader Interplay*, edited by H. Van den Bossche, pp. 537–548. Elsevier/North Holland, Amsterdam.
87. Trager, W., and Jensen, J. B. (1976): Human malaria parasites in continuous culture. *Science*, 193:673–675.
88. Trager, W., and Jensen, J. B. (1980): Cultivation of erythrocytic and exoerythrocytic stages of plasmodia. In: *Malaria*, Vol. 2, edited by J. P. Kreier, pp. 271–319. Academic Press, New York.
89. Undeen, A. H. (1975): Growth of *Nosema algerae* in pig kidney cell cultures. *J. Protozool.*, 22:107–110.
90. Viens, P., Lajeunesse, M. E., Richards, R., and Targett, G. A. T. (1977): *Trypanosoma musculi: in vitro* cultivation of blood forms in cell culture media. *Int. J. Parasitol.*, 7:109–111.
91. Wickham, J. M., Dennis, E. D., and Mitchell, G. H. (1980): Long-term cultivation of a simian malaria parasite *(Plasmodium knowlesi)* in a semi-automated apparatus. *Trans. R. Soc. Trop. Med. Hyg.*, 74:789–792.
92. Yamin, M. A. (1978): Axenic cultivation of the cellulolytic flagellate *Trichomitopsis termopsidis* (Cleveland) from the termite *Zootermopsis*. *J. Protozool.*, 25:535–538.

93. Yamin, M. A. (1980): Cellulose metabolism by the termite flagellate *Trichomitopsis termopsidis*. *Appl. Environ. Microbiol.*, 39:859–863.
94. Yamin, M. A. (1981): Cellulose metabolism by the flagellate *Trichonympha* from a termite is independent of endosymbiotic bacteria. *Science*, 211:58–59.

Molecular Biology of Parasites, edited by
J. Guardiola, L. Luzzatto, and W. Trager.
Raven Press, New York © 1983.

Genetics of Parasites

D. Walliker

Protozoan Genetics Unit, Institute of Animal Genetics, Edinburgh EH9 3JN, Scotland

Little is known of the genetics of most parasites, a state of affairs which is surprising in view of the fundamental importance of the subject in our understanding of the basic biology of other microorganisms. Knowledge of genetic mechanisms such as recombination and mutation in parasites is likely to have considerable practical as well as theoretical value. For example, an understanding of the genetic basis of drug resistance and antigenic variation should lead to the design of more rational methods of control than have been used hitherto.

The subject can be studied in two ways:

1. *Conventional genetic studies*, involving hybridization of strains and progeny analysis, for investigating the natural processes of genetic exchange and inheritance of specific characters in parasite populations.
2. *Molecular genetics*, for studying the basic organization of the parasite genome and the mechanism of gene expression.

The lack of progress in this field has been due to a variety of reasons, notably the difficulties of applying conventional techniques of genetic analysis to organisms with complex life cycles, the lack of suitable characters for use as genetic markers, and the lack of adequate methods of culture for most parasites. The purpose of this paper is to review the conventional genetic studies that have been carried out so far. For parasitic protozoa these studies have been restricted to organisms known to undergo a sexual phase in their life cycle, principally malaria parasites and coccidia. Among helminths, only schistosomes have been studied.

MALARIA PARASITES

Genetic studies on malaria parasites have been used to obtain information in the following areas:

1. The basic genetic organization of the parasite, especially the ploidy of blood forms.
2. The extent of genetic variation in enzymes in natural populations of parasites, with a view to identifying interbreeding groups.
3. Genetics of resistance to antimalarial drugs.
4. Genetics of virulence.

Most work has been carried out with the species of *Plasmodium* infecting rodents, especially *P. yoelii* and *P. chabaudi*. These have the advantage over other species of being easily maintained in the laboratory. Also available are a large number of wild isolates that provide a good source of natural genetic variation.

Genetic Techniques

The ideal method of hybridizing malaria parasites would be to mix purified male gametes (microgametes) of one strain with female gametes (macrogametes) of a second strain. However, reliable techniques for separating the two types of gamete are not yet available. Instead, crosses are made by making a mixture of male and female gametes of both lines. Assuming that random fertilization occurs, there will be an equal number of self- and cross-fertilization events, resulting in equal numbers of parent-type and hybrid zygotes.

In rodent malaria parasites, blood forms containing gametocytes are mixed together, either as a single infection in a rodent or in the chamber of a mosquito-feeding apparatus (Fig. 1). Mosquitoes are then fed on the mixture. The resulting zygotes are permitted to mature to sporozoites, which are used to infect further rodents. The blood forms that emerge in these animals are cloned, and each clone is examined for the characters distinguishing the parent lines. The presence of recombinant forms indicates that cross-fertilization of gametes has occurred.

This technique has been used to examine the inheritance of numerous characters in malaria, notably electrophoretic forms of enzymes, drug resistance, and virulence (1,17).

Genetics of Enzyme Variation

Enzyme forms, revealed by electrophoretic techniques, have proved of particular interest in numerous organisms because they are the products of genes. Variations in enzymes are thus a direct reflection of genetic differences.

Six enzymes have been investigated, using, principally, starch gel electrophoresis: glucose phosphate isomerase (GPI), 6-phosphogluconate dehydrogenase (PGD), lactate dehydrogenase (LDH), NADP-dependent glutamate dehydrogenase (GDH),

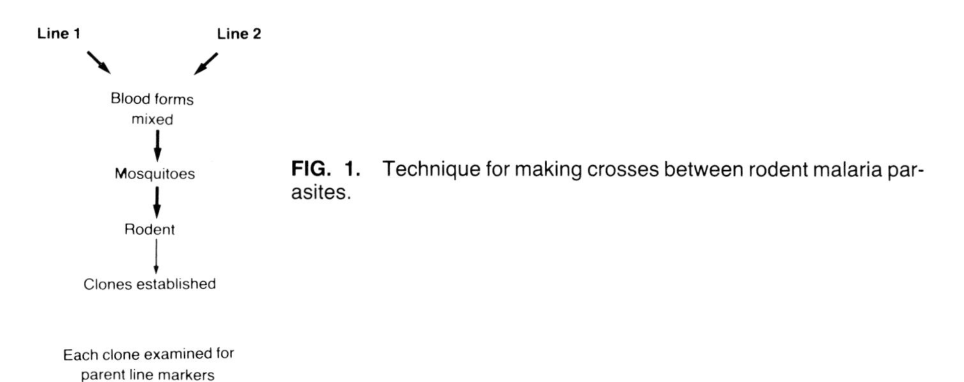

FIG. 1. Technique for making crosses between rodent malaria parasites.

adenosine deaminase (ADA), and peptidase E (PEPE). Considerable variation has been found among both rodent and human parasites (2).

Inheritance

The inheritance of enzyme forms can be illustrated by a cross between two *P. chabaudi* lines differing by two enzymes (Table 1). One line was characterized by forms of PGD and LDH denoted PGD-2 and LDH-3, respectively, and the other by PGD-3 and LDH-2. Following the cross, clones characterized by enzyme forms PGD-2, LDH-2, and PGD-3, LDH-3, as well as parental-type combinations, were isolated. Three principal conclusions could be drawn from this result:

1. Recombination between enzymes had occurred.
2. The blood forms were haploid. This was shown by the presence of only single forms of each enzyme in each progeny clone. If the blood forms were diploid, hybrid-type clones would exhibit both forms of each enzyme. It was clear, however, that segregation of each form of each enzyme had occurred before cloning, indicating that meiosis was taking place between zygote formation in the mosquito and the emergence of parasites into the blood.
3. Random mating of gametes had probably occurred. The numbers of recombinant and parental forms among the progeny (15:55) was close to the predicted proportions (1:3), taking account of self- as well as cross-fertilization events in the mosquito.

Enzyme Variation in Parasite Populations

Several studies have been made on the distribution of enzyme variants in natural populations of malaria parasites. Knowledge of the inheritance of such variants in crosses, as outlined above, makes it possible to determine whether the parasites in a given region belong to a single interbreeding population or whether they consist of more than one group of reproductively isolated organisms.

TABLE 1. *Recombination between forms of PGD and LDH in* Plasmodium chabaudi

Generation	Characteristics	Number of clones isolated
Parent lines		
Line 411AS	PGD-2, LDH-3	
Line 96AJ	PGD-3, LDH-2	
Progeny of cross		
Parental types	PGD-2, LDH-3	43
	PGD-3, LDH-2	12
Recombinants	PGD-2, LDH-2	9
	PGD-3, LDH-3	6

Among the rodent malaria parasites, the most detailed study on enzymes was carried out on a population in the Central African Republic (3). Two species were originally described in this region: *P. yoelii*, which preferentially invaded reticulocytes of its rodent host, and *P. chabaudi*, which invaded mature red cells. Isolates of *P. chabaudi* showed considerable enzyme diversity; when clones were made, however, the parasites could be classified into two distinct enzyme groups, one containing enzyme forms GPI-4, PGD-2 and -3, LDH-2, -3, -4 and -5, and GDH-5, and the other GPI-5 and -9, PGD-5, LDH-7, and GHD-6 (Table 2). It was clear that these two groups were reproductively isolated, even though mixed infections were common in wild-caught rodents. If hybridization between the two groups were taking place, parasite clones showing recombination between enzymes of each group would be found. Thus, two distinct species had been identified in these animals (subsequently named *P. chabaudi chabaudi* and *P. vinckei petteri*) where formerly only one (*"P. chabaudi"*) had been recognized.

Enzymes have also been valuable in distinguishing regional variation among parasites of a single species. Table 3 shows the enzyme forms of a selection of isolates of *P. yoelii* from different parts of Africa. In some instances the same enzyme forms are found in all regions, while in others they differ. Thus, similar forms of GPI, PGD, and LDH occur in all regions; *P. y. nigeriensis*, however, has a unique form of ADA, and each subspecies possesses a distinct form of GDH. Only the parasites of Cameroun and the Central African Republic are indistinguishable from one another. It seems probable that geographical isolation of these populations is leading to genetically divergent forms in each region. However, the

TABLE 2. *Enzyme forms of cloned isolates of rodent malaria parasites of the Central African Republic*

Species	Clone	GPI	PGD	LDH	GDH
P. chabaudi chabaudi	6AL	4	2	2	5
	3AR	4	2	3	5
	10AS	4	2	3	5
	2BJ	4	2	4	5
	57AF	4	2	5	5
	9AJ	4	3	2	5
	14AQ	4	3	2	5
	20CE	4	3	4	5
	2CW	4	3	4	5
	3CQ	4	3	5	5
P. vinckei petteri	1BS	9	5	7	6
	4BZ	9	5	7	6
	11CE	5	5	7	6
	2CR	5	5	7	6

Values for each enzyme refer to electrophoretic forms as detected by starch gel electrophoresis (see ref. 2).

TABLE 3. *Enzyme forms of some isolates of* P. yoelii *subspecies*

Species	Isolate	GPI	PGD	LDH	GDH	ADA
P. yoelii yoelii	17X	1	4	1	4	2
(Central African Republic)	33X	2	4	1	4	2
	86X	1	4	1	4	2
	5AD	1,2	4	1	4	2
	3AF	1	4	1	4	2
P. yoelii nigeriensis (Nigeria)	N67	2	4	1	2	1
P. yoelii killicki	193L	1	4	1	1	2
(Congo)	194ZZ	1	4	1	1	2
P. yoelii subsp.	EJ	1	4	1	4	2
(Cameroun)	EL	1	4	1	4	2

differences are not sufficiently large to prevent interbreeding of parasites when deliberate crosses are made in the laboratory. Crosses have recently been made between the *P. yoelii* parasites of each region, most of which indicate no preference for self- rather than cross-fertilization (ref. 7; and F. A. Lainson, personal communication).

In contrast to the rodent parasites, the human malaria parasite *P. falciparum* possesses similar enzyme forms in all countries examined so far (principally Africa and Southeast Asia), although some minor differences occur in the frequency of particular enzyme forms in different countries (12,16). Provisionally, it can be concluded that *P. falciparum* comprises a single population of potentially interbreeding organisms.

Genetics of Drug Resistance

Drug-resistant forms of *P. falciparum* are now widespread in many parts of the world, especially Southeast Asia and South America. Such resistance may be due to nongenetic physiological adaptations of the parasite to the drug or to genetic changes such as mutations, followed by selection by the drug. Work with rodent parasites has shown that genetic changes are an important cause of resistance to the two most commonly used antimalarials, chloroquine and pyrimethamine.

Pyrimethamine-resistant forms of *Plasmodium* can be obtained in the laboratory simply by exposing sensitive parasites to a single course of drug treatment. Such resistant forms may be obtained from cloned sensitive forms, and the resistance obtained is usually stable in the absence of the drug and following mosquito transmission. The level of resistance obtained may be higher than that of the drug dose used for selection. Each of these features suggests that gene mutations are involved in the resistance. This has been proved by examining the inheritance of the character of resistance in crosses between resistant and sensitive parasite lines. In *P. yoelii*, for example, a cross made between a pyrimethamine-resistant line characterized by GPI-1 and a sensitive line characterized by GPI-2 yielded recombinant forms (resistant, GPI-2 and sensitive, GPI-1) in the proportions expected if the resistance was due to a nuclear gene mutation (Table 4).

Gene mutation has also been shown to be the cause of low-level resistance to chloroquine in *P. chabaudi* (11). In this work, stable chloroquine resistance was obtained in a line already resistant to pyrimethamine by prolonged treatment with low doses of chloroquine (3 mg/kg). In a cross with a line sensitive to both drugs, recombination occurred between the chloroquine resistance and each of the other markers involved, including pyrimethamine resistance. It could be concluded, therefore, that the mutation causing chloroquine-resistance had occurred at a different locus from that causing pyrimethamine-resistance.

In contrast to pyrimethamine resistance, high-level resistance to chloroquine cannot be achieved by single-step treatment of sensitive organisms. Padua (8) was able to produce high-level chloroquine-resistance (30 mg/kg) in *P. chabaudi* by exposing parasites to gradual increases in drug concentration over many months. When this line was crossed with a drug-sensitive line, progeny clones were isolated that exhibited various levels of resistance from complete sensitivity to 30 mg/kg resistance (Fig. 2). The probable explanation of this result was that the high resistance had been produced by an accumulation of several mutant genes at different

TABLE 4. *Recombination between pyrimethamine-resistance and GPI-type in* Plasmodium yoelii

Generation	Characteristics	Number of clones isolated
Parent lines		
Line A	Resistant, GPI-1	
Line C	Sensitive, GPI-2	
Progeny of cross		
Parental types	Resistant, GPI-1	21
	Sensitive, GPI-2	30
Recombinants	Resistant, GPI-2	13
	Sensitive, GPI-1	7

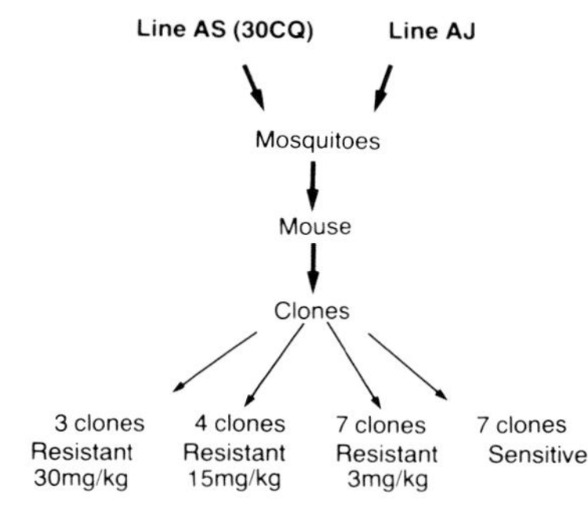

FIG. 2. Genetics of chloroquine-resistance in *P. chabaudi.* Line AS (30CQ) is resistant to chloroquine treatment of 30 mg/kg for 6 days. The resistance level of the progeny clones also refers to treatment for 6 days. Line AJ and sensitive clones are sensitive to chloroquine treatment of 3 mg/kg for 6 days.

SCHISTOSOMES

Schistosomes are the only helminths in which hybridization studies have been carried out. The technique for making crosses has been to infect snails with single parasites (miracidia) to obtain single-sex infections and then to expose mammal hosts with male and female cercariae derived from the parasite lines under investigation. The object of all such hybridization work so far has been to investigate whether the various schistosome species can interbreed. Taylor (15) carried out a series of crosses between members of the *Schistosoma haematobium*, *S. mansoni*, and *S. rodhaini* groups. Characteristics such as egg morphology, host infectivity, and enzymes have provided the principal markers used to test for hybridization.

REFERENCES

1. Beale, G. H., Carter, R., and Walliker, D. (1978): Genetics. In: *Rodent Malaria*, edited by R. Killick-Kendrick and W. Peters, pp. 213–245. Academic Press, New York.
2. Carter, R. (1978): Studies on enzyme variation in the murine malaria parasites. *Plasmodium berghei, P. yoelii, P. vinckei and P. chabaudi* by starch gel electrophoresis. *Parasitology*, 76:241–267.
3. Carter, R., and Walliker, D. (1975): New observations on the malaria parasites of rodents of the Central African Republic: *Plasmodium vinckei petteri* subsp. nov. and *Plasmodium chabaudi* Landau, 1965. *Ann. Trop. Med. Parasitol.*, 69:187–196.
4. Hoeijmakers, J. H. J., Frasch, A. C. C., Bernards, A., Borst, P., and Cross, G. A. M. (1980): Novel expression-linked copies of the genes for variant surface antigens in trypanosomes. *Nature*, 284:78–80.
5. Joyner, L. P., and Norton, C. C. (1975): Transferred drug-resistance in *Eimeria maxima*. *Parasitology*, 71:385–392.
6. Knowles, G., and Walliker, D. (1980): Variable expression of virulence in the rodent malaria parasite *Plasmodium yoelii yoelii*. *Parasitology*, 81:211–219.
7. Knowles, G., Sanderson, A., and Walliker, D. (1981): *Plasmodium yoelii:* a genetic analysis of crosses between two rodent malaria subspecies. *Exp. Parasitol.*, 52:243–247.
8. Padua, R. A. (1981): *Plasmodium chabaudi:* genetics of chloroquine-resistance. *Exp. Parasitol.*, 52:419–426.
9. Pfefferkorn, L. C., and Pfefferkorn, E. R. (1980): *Toxoplasma gondii:* genetic recombination between drug resistant mutants. *Exp. Parasitol.*, 50:305–316.
10. Rollinson, D., Joyner, L. P., and Norton, C. C. (1979): *Eimeria maxima:* the use of enzyme markers to detect the transfer of drug-resistance between lines. *Parasitology*, 78:361–367.
11. Rosario, V. E. (1976): Genetics of chloroquine-resistance in malaria parasites. *Nature*, 261:585–586.
12. Sanderson, A., Walliker, D., and Molez, J. F. (1981): Enzyme typing of *Plasmodium falciparum* from some African and other old world countries. *Trans. R. Soc. Trop. Med. Hyg.*, 75:263–267.
13. Shirley, M. W. (1978): Electrophoretic variation of enzymes: a further marker for genetic studies of *Eimeria*. *Z. Parasit.*, 57:83–87.
14. Tait, A. (1980): Evidence for diploidy and mating in trypanosomes. *Nature*, 287:536–538.
15. Taylor, M. G. (1970): Hybridisation experiments on five species of schistosomes. *J. Helminthol.*, 44:253–314.
16. Thaithong, S., Sueblinwong, T., and Beale, G. H. (1981): Enzyme typing of some isolates of *Plasmodium falciparum* from Thailand. *Trans. R. Soc. Trop. Med. Hyg.*, 75:268–270.
17. Walliker, D. (1976): Genetic factors in rodent malaria parasites and their effect on host-parasite relationships. In: *British Society for Parasitology Symposia, Vol. 14: Genetic Aspects of Host-Parasite Relationships*.
18. Walliker, D., Carter, R., and Morgan, S. (1973): Genetic recombination in *Plasmodium berghei*. *Parasitology*, 66:309–320.
19. Walliker, D., Sanderson, A., Yoeli, M., and Hargreaves, B. J. (1976): A genetic investigation of virulence in a rodent malaria parasite. *Parasitology*, 72:183–194.

Molecular Biology of Parasites, edited by
J. Guardiola, L. Luzzatto, and W. Trager.
Raven Press, New York © 1983.

Genetics and the Malarial Parasites

L. Luzzatto

International Institute of Genetics and Biophysics, CNR, 80125 Naples, Italy

Every living organism represents the product of its genome, the environment, and interactions between them. In the ideal case, the genome is the biological pole and the environment the physicochemical pole of the interaction process. However, in the real world no living organism develops without contact with other living entities, and the biological conditioning of development is therefore based not only in the genome, but also in the environment. If this is a truism for all organisms, such biological interactions are, by definition, closest in the case of parasitism: extreme, indeed, in the case of intracellular parasitism, wherein direct cell-to-cell interactions occur between cells with widely different genomes. The malarial *Plasmodia* and the human erythrocyte are a good example of this situation.

In principle, interactions could be expected to take place at various stages, such as invasion, intracellular development (schizogony), and release of merozoites. Therefore, genetically determined changes in the host cell that could affect the life cycle of the parasite might occur in membrane components or in any intracellular element of the erythrocyte and its metabolic machinery. In addition, we must consider genetic changes that, though not significantly altering the property of the red cell under normal conditions, might affect it once it is parasitized.

In practice, no particular genetic abnormality of human erythrocytes has been found to increase their susceptibility to malaria parasites, but several abnormalities have been found to decrease it. In the red cell membrane, if the substance responsible for Duffy antigen specificity is lacking, *P. vivax* infection is totally prevented. This has led to the concept that a Duffy-related substance is the "receptor" for this parasite. Recently, a similar suggestion has been made with respect to glycophorin, a major erythrocyte membrane protein carrying MN antigen specificity, and *P. falciparum*.

In the red cell cytoplasm, the major component is, of course, hemoglobin. There is ample evidence that the Hb S mutation exerts a protective effect with respect to *P. falciparum*, either because intracellular aggregates of deoxy-Hb S interfere with intracellular development of the parasite or because parasitized cells sickle and are removed by macrophages or because both these mechanisms operate. Among other cytoplasmic components, glucose-6-phosphate dehydrogenase is of special interest, since females heterozygous for this X-linked gene are genetic mosaics, and they have also been shown to have relative resistance against *P. falciparum*. The mechanism of this is currently being investigated by *in vitro* cultures. Other genetic abnormalities of human red cells that may affect susceptibility to malaria parasites include Hb C, Hb E, the thalassaemias, and ovalocytosis.

While red cell changes are especially apt to interfere with the intraerythrocytic stages of *Plasmodia*, it is important to keep in mind that other genetic host factors, though presently less amenable to experimental investigation, may be no less relevant. Thus, genetic variants of serum proteins might affect sporozoites or merozoites, as short as their free-living time may be. Tissue factors might affect preerythrocytic forms and preerythrocytic schizogony. HLA-linked genes certainly affect cellular and humoral immune response.

REFERENCES

1. Friedman, M. J., and Trager, W. (1981): Biochemistry of malaria parasites in human red blood cells. *Sci. Am.*, (3).
2. Livingstone, E. B. (1971): Malaria polymorpand human polymorphisms. *Annu. Rev. Genet.*, 5:33.
3. Luzzatto, L. (1979): Genetics of red cells and susceptibility to malaria. *Blood*, 54:961.
4. Miller, L. H., and Carter, R. (1976): Innate resistance in malaria. *Exp. Parasitol.*, 40:132.
5. Mourant, A. E., Kopec, A. C., and Domaniewska-Sobczak, K. (1978): *Blood Groups and Diseases*. Oxford University Press, Oxford.
6. Piazza, A., et al. (1973): HL-A variation in four Sardinian villages under differential selective pressure by malaria. In: *Histocompatibility Testing 1972*, edited by J. Dausset and J. Colombani, pp. 73–84. Munksgaard, Copenhagen.

Molecular Biology of Parasites, edited by
J. Guardiola, L. Luzzatto, and W. Trager.
Raven Press, New York © 1983.

Extrachromosomal Genetic Elements and Their Relation to Parasitism

John Guardiola

International Institute of Genetics and Biophysics, I80125 Naples, Italy

The informational content of a cell is rarely confined to chromosomal gene pools obeying to Mendelian genetic laws, but is also frequently conveyed by "non-Mendelian genetic compartments" composed of a number of heterogeneous informational molecules (or particles) that may coexist inside the same cell (2,6,14).

I wish to discuss here a possible role for the information carried by non-Mendelian genetic compartments (i.e., by extranuclear and extrachromosomal genetic elements) in conferring selective advantages to populations confronted with catastrophic stress situations, i.e., with sudden and dramatic changes of the environmental conditions such as the two below:

1. Sudden introduction of toxic compounds in the environment (for example, a bacterial population treated with antibiotics).
2. Sudden disappearance of species from a given environment (for example, rapid elimination, by the action of man or of a powerful pathogen, of the host of a very specialized parasite).

These are likely, in fact, to be two cases in which the population involved cannot adjust themselves to the new situation by means of evolutionary changes mediated by stochastic mutations. Any adaptation would, rather, require the "telescoping" of the required changes to shorter time scales.

DISTRIBUTION AND FUNCTION CODED FOR BY NON-MENDELIAN GENETIC COMPARTMENTS

Population geneticists have sophisticated analytical means (17) for determining the impact of the introduction of alleles transmitted in a Mendelian fashion on the genetic composition and fitness of populations confronted with selective pressures.

It is difficult to find in the literature a comparable analysis of the evolutionary role of genes belonging to non-Mendelian genetic compartments (4,17). Why this should be so can be justified neither on the basis of the rarity of such compartments nor on the supposed secondary importance of the functions coded for by the genes belonging to these compartments. In fact, nonchromosomal genetic compartments are nearly ubiquitous: some 50% of bacteria isolated in nature bear one plasmid, some more, and some also contain prophage DNAs (11,12). In eukaryotes the presence of mitochondria is universal, and all plants contain chloroplasts. Moreover, viruses, ds-RNA plasmids, plasmidlike DNAs, viroids, and obligate endosymbiotic organisms are widely distributed among these organisms (12).

A number of important functions is correlated with the presence of these elements, such as assimilatory functions (respiratory and photosynthetic ability, nitrogen fixation, ability to ferment unusual substrates), resistance mechanisms (to antibiotics, to toxic organic compounds, to divalent poisonous ions), and miscellaneous other properties (symbiotic capability, toxin production and resistance, sterility and incompatibility in sexual crosses).

The heterogeneity of the nonchromosomal genetic elements, the fact that they do not obey Mendelian laws, and the peculiarities of their mode of reproduction and transmission are thus the most likely reasons why the impact of these elements on evolution has long been neglected.

It will thus be enlightening to consider how host-parasite relationships and host-parasite coevolution are influenced by relevant traits specified by genes of non-Mendelian genetic pools.

EXTRANUCLEAR AND EXTRACHROMOSOMAL DETERMINANTS AND THEIR RELATION TO PATHOGENICITY AND PARASITISM

Some examples of the extrachromosomal determination of functions relevant to parasitic relationships or to pathogenicity are known. The role of *Agrobacterium tumefaciens* Ti plasmid in the genetic colonization of plant cells is particularly well known (3). Some evidence is also available that the phytopathogenicity of some bacterial strains might be plasmid-coded (5,10). *Escherichia coli* strains pathogenic to man need to adhere to the intestinal tissue, and this ability is a plasmid-determined function (9). Some races of the fungal parasite *Helminthosporium maydis* are pathogenic only to *Zea mays* strains carrying the so-called T (Texas) cytoplasm (7). T cytoplasm is characterized an an abnormal architecture of mitochondrial DNA. The specificity of some *Puccinia graminis* races is also cytoplasmically inherited (8). The susceptibility or resistance of the mosquito complex *Aedes scutellaris* to the filarial parasites *Brugia malayi* and *B. pahangi* is maternally inherited, i.e., is not transmitted according to Mendelian laws (16).

PROKARYOTIC PLASMIDS

Our knowledge of the biology and genetics of parkaryotic extrachromosomal elements is by far the most advanced with respect to the subject of this chapter, although not directly in connection with parasitology (11). Some general conclusions can thus be drawn.

The extrachromosomal elements present in prokaryotes are essentially of two types: plasmids and phages. I consider here mainly plasmids. Prokaryotic plasmids are in essence autonomously replicating circular DNA molecules that exploit the

TABLE 1. *Genera able to accept naturally occurring plasmids isolated from Enterobacteriaceae and Pseudomonadaceae*

Acinetobacter, Aeromonas, Agrobacterium, Azotobacter, Chromobacterium, Citrobacter, Enterobacter, Erwinia, Escherichia, Klebsiella,	*Proteus, Providencia, Pseudomonas, Rhizobium, Rhodopseudomonas, Rhodospirillum, Salmonella, Shigella, Serratia, Vibrio, Yersinia.*

(Adapted from ref. 12, with permission.)

resources of the host cell for their maintenance and replication. Some basic properties of plasmids are of importance in this context. First, we consider plasmid transmissibility. Plasmids are, in fact, subdivided into self-transmissible and nontransmissible types. Self-transmissible plasmids possess transfer mechanisms that endow them with the ability to invade plasmid-free cells by conjugation. The transfer is infective, and a plasmid-free population can be converted very rapidly to a plasmid-bearing population by a few original cells carrying the plasmid. Nontransmissible plasmids, although devoid of transfer mechanisms of their own, can eventually be transmitted. This may happen by one of three different means: exploitation of the transfer ability of a coresident self-transmissible plasmid; transduction, whereby the plasmid is picked up by a phage and spread in the population; and spontaneous transformation, in which the host cell lyses accidently or as a result of senescence.

The second relevant property of plasmids is their ability to pick up genetic material from a coresident plasmid or from a chromosome, leading to new association of functions and, more important, allowing for mobilization of genetic information (12).

Of fundamental importance is the ability of many plasmids to transfer genetic information not only among individuals of the same species, but also from cells of a given bacterial species to cells of a distantly related one, so that advantageous properties need not be evolved *de novo* any time they are needed, provided that a channel for genetic interchange, such as plasmids and phages, is available (12).

The ability of different genera to accept plasmids from *Enterobacteriaceae* and *Pseudomonadaceae* is exemplified in Table 1. Some of the functions carried by plasmids, and thus susceptible to intergeneric transfer, are reported in Table 2. The versatility of plasmids in acting as carriers of genetic information is enhanced by their ability to merge with other replicons so as to generate new plasmids possessing additional favorable properties.

IMPACT OF PLASMIDS ON EVOLUTION OF ANTIBIOTIC-RESISTANT POPULATIONS

One example of the above-mentioned impact on evolution resulting from the properties of plasmid relates to the spread of plasmid-coded antibiotic resistance under selective conditions.

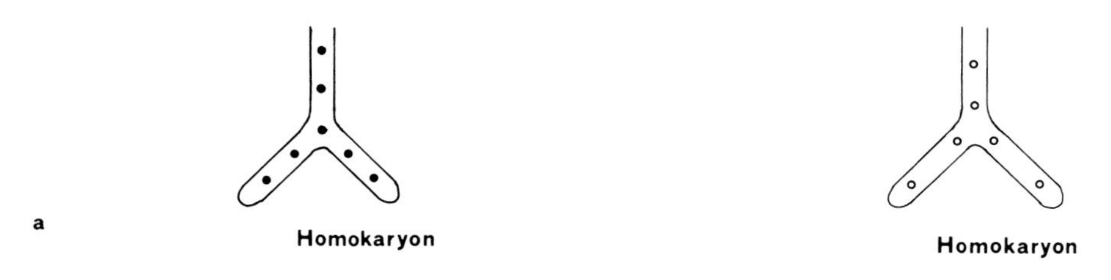

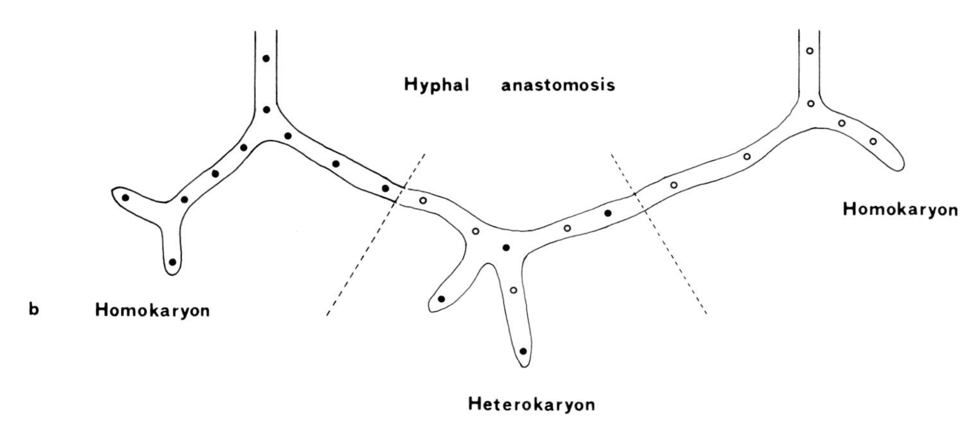

FIG. 1. Hyphal fusion with exchange of genetic material in filamentous fungi. (Data from ref. 6.)

TABLE 2. *Some functions carried by prokaryotic plasmids*

Resistance genes
Resistance to penicillin, streptomycin, neomycin, chloramphenicol, erythromycin, kanamycin, arsenate, arsenite, lead, bismuth, nickel, cadmium, mercury, colicin, phages, ultraviolet light
Steps in metabolism
Degradation of camphor, oatane, salicylate, naphtalene; degradation of stable RNA; DNA polymerase; use of sucrose and lactose
Enzymes for restriction and modification
R·Eco·R1 endonuclease, DNA methylase specific for cytosine
Miscellaneous
Colicinogeny, phage modification, conjugal transmissibility, sensitivity to male-specific phages, surface exclusion, sensitivity to silicate, sodium dodecyl sulfate sensitivity, α-hemolysin production, enterotoxin production, K88 surface antigen production

(Adapted from ref. 12, with permission.)

It is commonly known that determinants for resistance to a large number of antibiotics are carried by plasmids (1). The occurrence of plasmids coding for antibiotic resistance in bacterial species previously devoided of them parallels the use of given antibiotics in medical and veterinary practice (1,13). It is conceivable

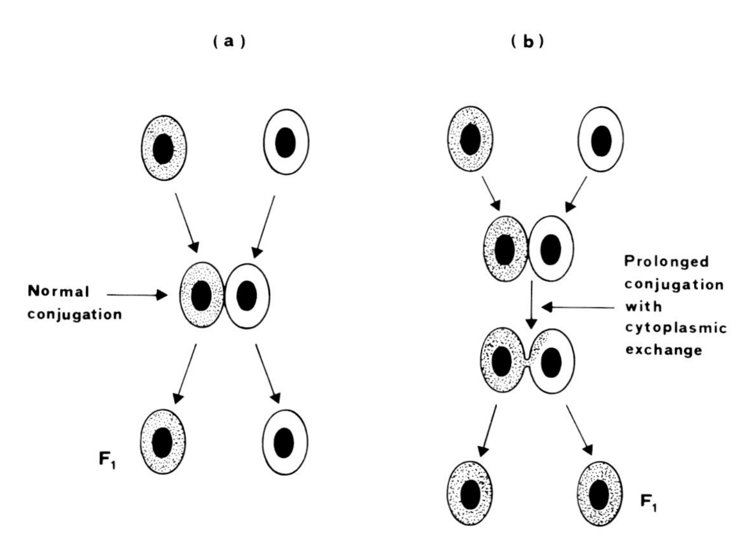

FIG. 2. Normal conjugation and prolonged conjugation with cytoplasmic exchange in *Paramecium.* (Data from ref. 2.)

that a given antibiotic-resistance-determining plasmid might spread from a harmless reservoir species to a pathogenic bacterial population; under selective pressure, the pathogenic cells that have inherited the plasmid will rapidly displace the plasmid-free population (1,13,15).

A comparison between the modes of antibiotic resistance mediated by plasmid determinants and antibiotic resistance obtained through mutational alteration of chromosomal genes was discussed by Reanny (12), who concluded that the speed of adaptation in natural ecosystems determined by the properties of extrachromosomal elements should largely exceed that which obtains under natural selection of spontaneous favorable mutations in chromosomal *loci*.

PLASMIDS AND PARASITISM: A HYPOTHETICAL CASE

One might wish to extend the suggestions stemming from the epidemiological analysis of antibiotic resistance to a hypothetical ecological niche containing two hosts, A and B, each with a highly specialized bacterial parasite, say α and β, where α is able to parasitize only A and β is able to parasitize only B. Let us assume, safely enough, that the determinants that enable the two bacterial strains to recognize only their specific host depend on genes or sets of genes that are not allelic and are thus completely different. If by accident host B, for example, is suddenly wiped out by some action of man or by a powerful facultative pathogen also existing in this hypothetical niche, the bacterial population will by no means be able to survive, since it is unable to parasitize A; neither is it reasonable to expect that there will be enough time to evolve a new and complex specificity mechanism.

On the other hand, if we formally consider specificity mechanisms such as the determinants of antibiotic resistance, we can expect that the ability to parasitize A might be transferred from strain α to strain β in the same way that resistance to an antibiotic is transferred from a reservoir strain to a pathogenic strain by means of a broad-host-range plasmid.

CONCLUSIONS

It is possible that analogous intergeneric exchanges take place among different eukaryotic organisms? In fact, such a possibility requires not only the existence of extranuclear genetic elements, which, as previously mentioned, certainly exist among eukaryotes, but also demands the means to accomplish such genetic exchanges.

Indeed, at least in lower eukaryotes we know of processes that might be exploited in order to accomplish that goal. For example, we know that fungal hyphae fuse (anastomose, see Fig. 1) so that a physical bridge between two different cytoplasms is formed, and we also know of mating events among protozoa in which different cytoplasm comes into contact (see Fig. 2).

The existence of carriers of extranuclear genetic information and the opportunity for their transfer suggest that a horizontal spread of genetic information might also take place across interspecific and intergeneric barriers—at least among lower eucaryotes. The exchange of pre-evolved and "pretested" information packages might thus be a general feature by which many organisms achieve a rapid adaptative response when needed.

REFERENCES

1. Barber, M. (1963): Development of drug resistance by *Staphylococci* in *vitro* and in *vivo*. *Int. Rev. Cytol.*, 14:267–279.
2. Beale, G., and Knowles, J. (1978): *Extranuclear Genetics*. Edward Arnold, London.
3. Chilton, M. D., Drummond, M. H., Merlo, D. J., Sciaky, D., Montoya, A. L., Gordon, M. P., and Nester, E. W. (1977): Stable incorporation of plasmid DNA into higher plant cells: The molecular basis of crown gall tumorigenesis. *Cell*, 11:263–271.
4. Cosmidas, L. M., and Tooby, J. (1981): Cytoplasmic inheritance and intragenomic conflict. *J. Theor. Biol.*, 89:83–129.
5. Garibaldi, A., and Gibbins, L. N. (1975): Induction of avirulent variants in *Erwinia stewartii* by incubation at supra-optimal temperatures. *Can. J. Micribiol.*, 21:1282–1287.
6. Jinks, J. L. (1964): *Extrachromosomal Inheritance*. Prentice-Hall, Englewood Cliffs, N.J.
7. Johnson, R. (1976): Development and use of some genetically controlled lines for studies of host-parasite interactions. In: *Biochemical Aspects of Plant-Parasite Relationships*, edited by J. Friend and D. R. Threlfall. Academic Press, London.
8. Johnson, T. (1946): Variation an inheritance of certain characters in rust fungi. *Cold Spring Harbor Symp. Quant. Biol.*, 11:85–93.
9. Kehoe, M., Sellwood, R., Shipley, P., and Dougan, G. (1981): Genetic analysis of K-88-mediated adhesion of enterotoxinogenic *Escherichia coli*. *Nature*, 291:122–126.
10. Lacy, G. H., and Leary, J. V. (1979): Genetic systems in phytopathogenic bacteria. *Annu. Rev. Phytopathol.*, 17:181–202.
11. Lewin, B. (1977): Plasmids and phages. *Gene Expression*, Vol. 3. John Wiley, New York.
12. Reanny, D. (1976): Extrachromosomal elements as possible agents of adaptation and development. *Bacteriol. Rev.*, 40:552–590.
13. Richmond, A. S., Simberkoff, M. S., Rahal, J. J., Jr., and Schaeffer, S. (1975): R factors in gentamycin-resistant *S. aureus* in three hospitals. *Lancet*, 2:1176–1178.

14. Sager, R. (1972): *Cytoplasmic Genes and Organelles*. Academic Press, London.
15. Speller, D. C. E., Stephens, M., Raghunath, D., Viant, A. C., Reeves, D. S., Broughall, J. M., Wilkinson, P. J., and Holt, H. A. (1976): Epidemic infection by a gentamycin resistant *S. aureus* in three hospitals. *Lancet*, 1:464–466.
16. Trpis, M., Duherkopf, R. E., and Parker, K. L. (1981): Non-mendelian inheritance of mosquito susceptibility to infection with *Brugia malayi* and *Brugia pahangi*. *Science*, 211:1435–1437.
17. Wright, S. (1968): *Evolution and the Genetics of Populations*. The University of Chicago Press, Chicago.

Molecular Biology of Parasites, edited by
J. Guardiola, L. Luzzatto, and W. Trager.
Raven Press, New York © 1983.

Caenorhabditis elegans: A Model System for the Study of Nematodes

Paolo Bazzicalupo

International Institute of Genetics and Biophysics, 80125 Naples, Italy

Nematodes are a very large group of animals. The estimated 500,000 species represent an independent phylum, and a very successful one, since they are found, with the exception of the pelagic and aerial habitats, in every type of environment. The great majority of nematodes are free-living and inhabit in large numbers the top few centimeters of the ocean's bed, fresh water muds, and a variety of soils. In the soil, where it has been measured, their biomass is comparable to that of insects (Kuhnelt et al., 1976). A few hundred species are extremely important in human health and agriculture because of their parasitic relationship to plants and animals. In humans, parasitic nematodes can cause very severe diseases, such as filariasis and river blindness *(Oncocercus)*. Other human parasites, such as *Ascaris*, although they pose a less severe threat to the welfare of the affected individual, are important because of the extremely large number of people that are infested with them. Crop losses due to plant parasitic nematodes are estimated at hundreds of billions of dollars per year worldwide. Because of their economic importance, parasitic nematodes have been studied more intensely than other members of this group, but the difficulties of working with parasites in the laboratory and the concentration on the specific damage that a particular parasite inflicts on its host have left most aspects of their biology unexplored. Thus, surprisingly little is known about the physiology, biochemistry, anatomy, and behavior of parasitic nematodes, especially in terms of modern biological concepts. The need for a better understanding of the basic biology of nematodes is particularly felt at this time. From this perspective, then, the great interest offered by the free-living species *Caenorhabditis elegans* derives from the fact that it is the only nematode for which there is a growing body of knowledge at the cellular, genetic, and molecular levels. This is the result of increased basic research with this species.

In the past several years many researchers have been attracted by *C. elegans* as a model system because of the advantages that its small size, limited number of cells, short life cycle, and mode of reproduction offer for its study in the laboratory. For reasons that will be apparent, it is an ideal organism in which to attempt the integration of high-resolution genetic, biochemical, and morphological analysis necessary to the study of complex biological phenomena such as development and behavior.

The interest of parasitologists in the work done on this species is further justified when one considers the anatomical and physiological similarities among nematode species, an aspect that has led to the comment that "the elementary student may be forgiven at times for thinking that there is only one nematode but that the model comes in different sizes and a great variety of life histories" (Harris and Crofton, 1957). Naturally, the study of free-living *C. elegans* cannot deal with parasitism directly, and it cannot substitute for a system suitable for the study, in the laboratory, of the parasitic nematodes. Still, an understanding of as many aspects as possible of the biology of one nematode and the application of modern biological techniques to its study should be helpful in approaching some of the problems one faces when dealing with nematode infections in animals and plants.

The increased interest in *C. elegans* has produced a variety of meaningful results, but a thorough report on the state of these studies is outside the scope of this article. Extensive reviews of most aspects of the work on *C. elegans* have appeared (Zuckerman, 1980; Croll, 1976). Rather, the aim of this article is to give the reader interested in the molecular biology of parasites a general description of the biology of an intensely studied nematode and some examples of the main approaches followed in such studies.

BIOLOGY OF *C. elegans*

Gross Anatomy

Like all other nematodes, *C. elegans* is spindle-shaped, unsegmented, and bilaterally symmetrical. Figures 1 and 2 outline the general anatomy. The adult worm is about 1 mm long and about 100 μm in diameter. It is transparent, and it has therefore been possible to study the cellular anatomy of live animals with high-resolution optical microscopy. Because of its small size, complete serial sectioning of individuals for electron microscopy has not been a prohibitive task.

Its body is lined with a tough elastic noncellular cuticle made in part of crosslinked collagen. The cuticle is secreted by an underlining hypodermal syncytium, which also delimits the body cavity of nematodes. This cavity, also called the pseudocoelom, is filled with liquid under positive pressure, which is important for locomotion and feeding. The pseudocoelomic fluid bathes all the internal organs so that they receive from it the nutrients absorbed through the intestine and the oxygen diffusing in from the environment. Cuticle, hypodermal syncytium, and muscle cells make up the body wall of the worm. There are four rows of muscle cells that run anteroposteriorly inside the hypodermis in the subventral and subdorsal positions; the body wall muscles are used for locomotion. More than half the nerve cells of the organism are in the head region in ganglia associated with the nerve ring. This is located around the pharynx between the two bulbs. The nervous system includes a dorsal nerve cord that does not contain cellular elements, a ventral nerve cord containing motor neurons, and some small ganglia associated with the ventral cord and with sexual structures. The alimentary canal is formed by a mouth that

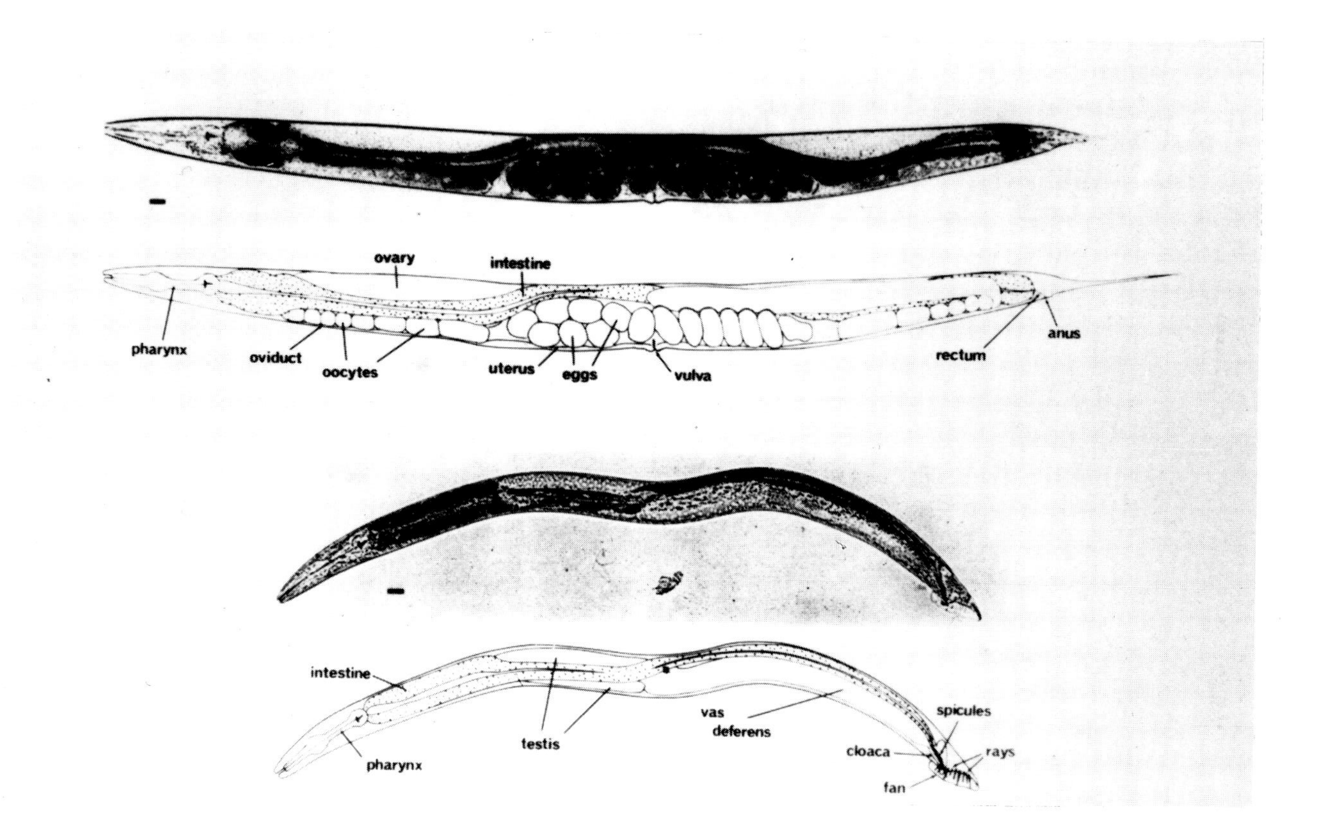

FIG. 1. Adult hermaphrodite (above) and male (below), lateral views; bright-field illumination. ×137. Bar = 20 μm. (From Sulston and Horvitz, with permission.)

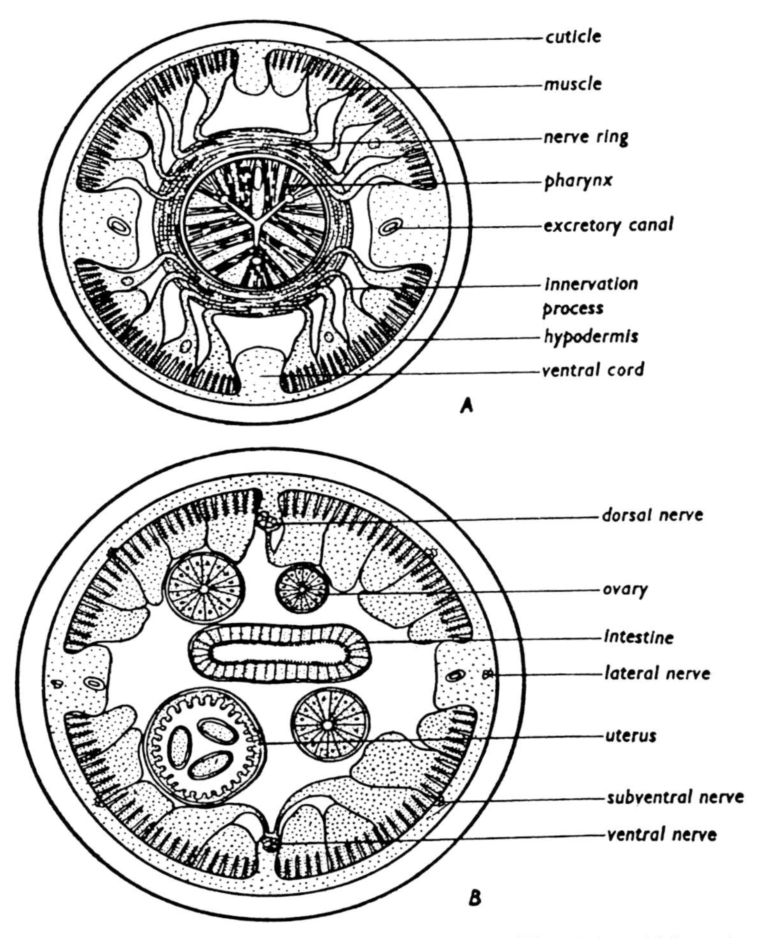

FIG. 2. Transverse sections through the pharyngeal region **(A)** and the middle region **(B)** of a nematode. (From Lee, 1965, with permission.)

leads to a muscular and glandular pharynx. The pharynx has two bulbs and a triradiate structure and pumps liquid and bacteria into the intestine, which is a straight tube lined with microvilli that extends posteriorly and terminates in a rectum and anus.

There are two sexes of *C. elegans*: self-fertilizing hermaphrodites and males. Most other nematodes differ from *C. elegans* in this respect, having a male and a female sex. The reproductive apparatus of the hermaphrodites is composed of two reflexed gonadal arms meeting at a midventral vulva through which fertilized eggs are laid and males can inject their sperm. Each arm of the gonad consists of an ovary with oviduct leading into a spermatheca and a uterus. Oocytes are produced in the ovary and reach maturity moving in the oviduct toward the spermatheca, where they are fertilized by the sperm stored there. The fertilized egg passes into the uterus, where it starts to cleave and is then laid outside the body. In the male, sperm is formed in a single testis terminating in the vas deferens. The vas deferens extends posteriorly and connects with the rectum, forming the cloaca at the tail end of the animal. The copulatory apparatus is formed by a bursa with spicules, rays, and a fan.

Life Cycle

The natural habitat of *C. elegans* is the soil, where it feeds on bacteria and small particles of decaying matter. In the laboratory it is usually grown between 18 and 23°C on solid or liquid media seeded with *Escherichia coli* (Brenner, 1974). Dougherty et al. (1959) have also described methods for its axenic cultivation. *C. elegans* usually reproduces as self-fertilizing hermaphrodites, which do not mate with each other. They have five pairs of autosomes and one pair of X chromosomes. Males appear in hermaphrodite populations by spontaneous events of meiotic nondisjunction at a frequency of about 0.3% and possess only one X chromosome in addition to the autosomal complement (Brenner, 1974; Hodgkin et al., 1979). Hermaphrodites reach sexual maturity in about three days and will produce by self-fertilization about 300 progeny that are virtually all hermaphrodites; if mated with males, one hermaphrodite will produce over a thousand cross progeny, 50% of which are males.

The fertilized egg is laid by the female at about the 30-cell stage; it is surrounded by an eggshell that is impermeable to most solutes and protects the embryo from the environment. Embryogenesis is complete in about 14 hr at 20°C. During the first 6 hr, the cleavage phase, about 500 cells are formed, and some important but limited cell migrations occur. During the second half of embryonic development, differentiation and morphogenesis are the important processes. They lead to the formation of a first-stage larva (L1) that starts to move inside the shell and then hatches by virtue of mechanical forces and weakening of the shell through enzymatic digestion. There is no increase in the volume of the animal during embryogenesis. The first-stage larva is 400 μm long and on hatching will immediately start to crawl around and feed. As it grows, the larva undergoes four moults before reaching sexual maturity. During postembryonic development, sexual differentiation takes place with formation of the germ cells, the gonad, and the secondary sexual structures. In addition, some of the cells present at hatching will divide, forming new motor neurons and more muscle and hypodermal cells. The excretory system, the anterior sensory nervous system, and the pharynx are complete at hatching (Sulston and Horvitz, 1977).

At hatching the gonad primordium is formed by four cells, two of which give rise to germ cells and two to the somatic structures of the gonad. In the hermaphrodite the gonad first produces sperm (about 300), storing them in the spermatheca; it then produces oocytes. The hermaphrodite has more than 1,500 oocytes, all of which can be fertilized by male sperm. But the brood size of the unmated hermaphrodite is limited to about 300 by the number of sperm it produces.

CELL LINEAGES AND DEVELOPMENTAL STUDIES

One of the main areas of research about *C. elegans*, and the one that probably has produced the most impressive results, is the detailed anatomical and developmental study of the organism at the cellular level. Thanks to the accessibility of live worms at all stages to microscopic analysis, and to the effort of several research groups, a complete cell-by-cell description of the anatomy and development of *C. elegans* from the one-cell stage to the adult is now available (Ward et al., 1975; Albertson and Thomas, 1976; Sulston, 1976; White et al., 1976; Sulston and Horvitz, 1977; Deppe et al., 1978; Kimble and Hirsh, 1979; and Sulston, personal communication, 1981).

The cellular anatomy and development of *C. elegans* are strictly determined and invariant: the number, the position, the function, and the lineage history of the cells are the same for different individuals of the same genotype and developmental stage. This reproducibility has allowed the reconstruction of the timing, orientation, and relative size of daughters of all the cell divisions. It is known when and where cell migration occurs, which cells die, and when gross morphological differentiation occurs. The number, position, and lineage of the cells of every structure of the adult worm are known (Tables 1 and 2, Figs. 3 and 4). The major exception to this rule of lineage invariance is given by the descendants of the P4 cell, which will form the germ line. Germ cells always derive from P4, but in different animals the cell divisions that produce the thousand and more gametes are different.

Sulston compiled the results of many groups and of his own work, and at the Third International Meeting on *C. elegans*, May 6–May 10, 1981, Cold Spring Harbor Laboratory, he presented a complete lineage chart for *C. elegans*, excluding the germ line. This result, unmatched for multicellular organisms, is important because it can be used as a precise reference for questions about development. The effects of various manipulations on the developing animal have been compared at the cellular level with the normal course in an effort to define the rules governing development. Thus, experiments have been done to assess the regulative capacities of cells during development or to distinguish between phenomena that are dependent on cell lineage or on the effect of position and interaction with other cells. For instance, laser ablation of some cells during postembryonic development has generally shown the lack of a regulative capacity in this organism. There are a few cells, though, that are capable of partial compensation for the lack of a neighboring cell; by and large, however, they will do so by changing the pattern of their division to that of the cell for which they are substituting (Sulston and Horvitz, 1977; Kimble et al., 1979; Sulton and White, 1981).

This same principle—that differentiation is strictly determined by lineage history and that regulation is exerted by activation of specific patterns of cell divisions—is consistent not only with the observation of normal development, but also seems to be confirmed by the analysis of lineage and cell differentiation in some mutants. In some developmental mutants the defect is due either to changes in the pattern of divisions of particular cells or to a typical change in the division pattern of many

TABLE 1. *Nuclear counts, hermaphrodite*

	Present in young L1		Derived from postembryonic lineages		
	Nondividing nuclei	Blast cells	Surviving nuclei	Cell deaths	Present in adult
Lateral hypodermis					
Seam	2	—	30	—	32
Syncytial	20	—	98	—	118
Neuronal or glial	8	—	28	—	36
Total	30	20 (H,V,T,Q)	156	8	186
Ventral cord and associated ganglia					
Neuronal or glial	33	—	56	9	89
Hypodermal	0	—	12	1	12
Vulva	0	—	22	0	22
Total	33	13 (P)	90	10	123
Mesoderm					
Body muscles	81	—	14	0	95
Sex muscles	0	—	16	0	16
Coelomocytes	4	—	2	0	6
Digestive tract muscles	4	—	0	0	4
Head mesodermal cell	1	—	0	0	1
Total	90	1 (M)	32	0	122
Intestine	6[a]	14[a] (I)	28	0	34
Head					
Neuronal, glial, small structural	201[b]	—	4	0	205[b]
Hypodermal (dorsal, ventral)	15	—	1	0	16
Pharynx	80	—	0	0	80
Pharyngeal-intestinal valve	6	—	0	0	6
Excretory system	4	—	0	0	4
Total	306	2 (G)	5	0	311
Tail					
Neuronal or glial	19	—	1	0	20
Hypodermal	7	—	1	0	8
Rectal glands	3	—	0	0	3
Total	29	1 (K)	2	0	31
Other tail ectoderm (B,C,E,F)	4	0	0	0	4
Total, excluding gonad	498	51	313	18	811

[a]The L1 intestine contains 20 nuclei. Of these, 6 never divide; in a given individual, 10–14 of the others divide.
[b]Estimate.
From Sulston and Horvitz (1977), with permission.

TABLE 2. *Nuclear counts, male*

	Present in young L1		Derived from postembryonic lineages		
	Nondividing nuclei	Blast cells	Surviving nuclei	Cell deaths	Present in adult
Lateral hypodermis					
Seam	2	—	36	—	38
Syncytial	20	—	104	—	124
Neuronal or glial	6	—	28	—	34
Rays	0	—	54	—	54
Total	28	20 (H,V,T,Q)	222	26	250
Ventral cord and associated ganglia					
Neuronal or glial	33	—	70	4	103
Hypodermal	0	—	10	1	10
Unknown	0	—	16	0	16
Total	33	13 (P)	96	5	129
Mesoderm					
Body muscles	81	—	14	0	95
Sex muscles	0	—			
Coelomocytes	4	—	42	0	46
Unknown	0	—			
Digestive tract muscles	4	—	0	0	4
Head mesodermal cell	1	—	0	0	1
Total	90	1 (M)	56	0	146

TABLE 2. *(continued)*

	Present in young L1		Derived from postembryonic lineages		
	Nondividing nuclei	Blast cells	Surviving nuclei	Cell deaths	Present in adult
Intestine	6[a]	14[a] (I)	28	0	34
Head					
Neuronal, glial, small structural	205[b]	—	4	0	209[b]
Hypodermal (dorsal, ventral)	15	—	1	0	16
Pharynx	80	—	0	0	80
Pharyngeal-intestinal valve	6	—	0	0	6
Excretory system	4	—	0	0	4
Total	310	2 (G)	5	0	315
Tail					
Neuronal or glial	19	—	1	0	20
Hypodermal	7	—	1	0	8
Rectal glands	3	—	0	0	3
Total	29	1 (K)	2	0	31
Other tail ectoderm	0	4 (B,C,E,F)	66	5	66
Total, excluding gonad	496	55	475	36	971

[a]The L1 intestine contains 20 nuclei. Of these, 6 never divide; in a given individual, 10-14 of the others divide.
[b]Estimate.
From Sulston and Horvitz (1977), with permission.

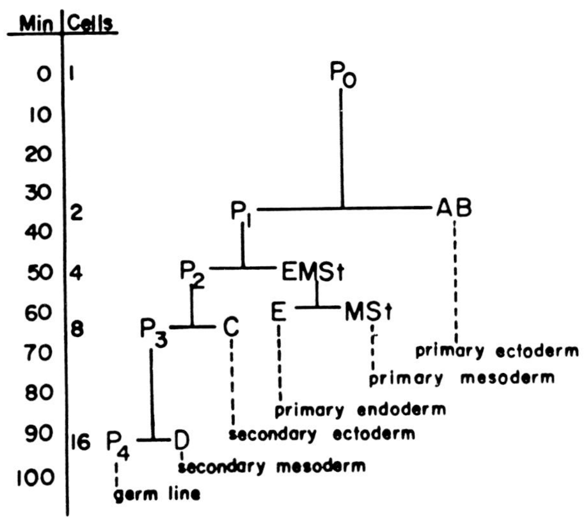

FIG. 3. Early cleavages in *C. elegans*. Formation of the precursors of the somatic tissue (AB, MSt, E, C, and D) and of the germ line precursor P4. The vertical axis indicates time in minutes after fertilization and total number of cells in the embryo at each time.

cells (Horvitz and Sulston, 1980; Sulston and Horvitz, 1981; and M. Chalfie et al., personal communication, 1981). The study of the role of genes in development, which is a major unsolved question in modern biology, seems particularly promising in this system, which is amenable to genetic manipulation and has a strictly determined development that is now known in great detail.

GENETIC APPROACHES

Another major approach followed in the study of *C. elegans* has been genetic analysis. *C. elegans* is an ideal system in which to use this approach because it is small and has a short reproductive cycle (three days at 20°C) with large brood size (300 to 1,000 progeny/worm); these factors make obtaining and handling large numbers of animals faster and easier than in other multicellular organisms. It has two modes of reproduction: the mating of males with hermaphrodites allows genetic crosses to occur, whereas self-fertilization by the hermaphrodite is advantageous in that homozygous animals appear spontaneously one generation after heterozygosity is established. This simplifies the isolation of clones of isogenic animals and favors detection of recessive mutations. Finally, *C. elegans* can be revived after storage in liquid nitrogen, which makes the maintenance of large collections of mutant strains a relatively simple task.

Genetic analysis has been used to study virtually every aspect of the biology of this organism. It is therefore impossible to review here all the genetic studies. Following the basic studies by Brenner (1974), a significant amount of work aimed

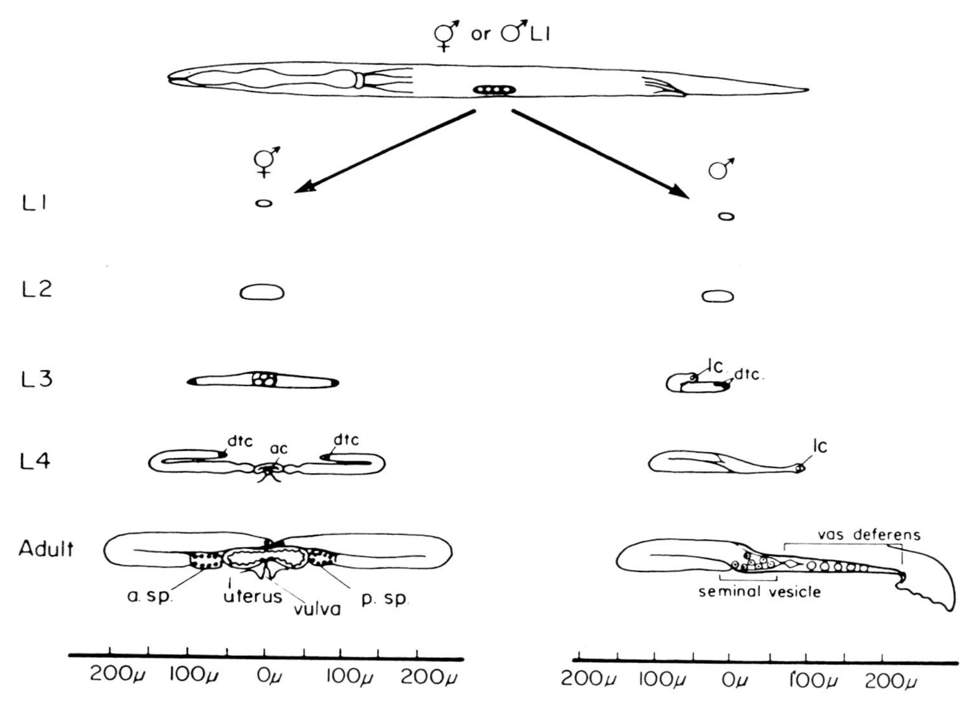

FIG. 4. Overview of gonadal development in hermaphrodites and males. At the top, the four-celled gonadal primordium is shown in its midventral position in the newly hatched worm. In hermaphrodites (left column), the developing gonad elongates anteriorly and posteriorly during L1, L2, and L3. The growing tips reflex around the time of the L3–L4 moult. In males (right column), the developing gonad initially grows only in the anterior direction and reflexes at about the L2–L3 molt. The main somatic structures of the adult hermaphrodite and male are shown schematically in the bottom two drawings. dtc, distal tip cell; ac, anchor cell; lc, linker cell; a.sp. anterior spermatheca; p.sp., posterior spermatheca. In the hermaphrodite, the adult gonad consists of 143 somatic cells and contains 300 sperm (150/arm) and 1000–1500 oocyte nuclei. In the male, the adult gonad consists of 56 somatic cells and 1500 sperm. (From Kimble and Hirsh, 1979, with permission.)

at defining and improving the genetic analysis itself. Strategies for mutagenesis, identification of mutants, mapping mutations, and defining genes by complementation analysis have been refined. The power of genetic manipulations has been increased by the isolation of duplications, deficiencies, chromosomal rearrangements, and informational suppressors (reviewed by Herman and Horvitz, 1980). A genetic map has been constructed, and more than 300 genes have been located on the six linkage groups.

Thousands of mutants have been isolated; I group them here as morphological, behavioral, developmental, and biochemical types—a division that is for the purpose of exposition only. One should bear in mind that there is some overlap and that assignment to groups is sometimes arbitrary.

Morphological Mutants

Morphological mutations are useful for genetic studies because they provide visible markers. "Small," "long," "dumpy," and "blistered" are some of the descriptive names of these mutants. It is worth mentioning that the phenotypes of blistered, rollers, and at least some of the dumpy mutants are due to alterations in the structure of the cuticle of the worm. As mentioned, the cuticle contains a crosslinked collagen. It is hoped that the combination of thorough mutational analysis with electron microscopy, protein chemistry, and cloning of collagen genes may help in elucidating the structure of this organ, which seems to be of fundamental importance in the physiology and survival of all nematodes, especially of parasitic species.

Behavioral Mutants

Behavioral mutants include those affected in their ability to move in general, and in their response to various stimuli, including temperature, chemicals, and touch. The most easily observable behavior of *C. elegans* in the laboratory is its movement on the surface of the Petri plates on which the worms are usually grown. The wild type worm moves with a smooth sinusoidal wave; it is therefore relatively easy to spot mutant worms that either do not move at all or move in abnormal ways. Many mutations—revealed in a variety of defects, including paralysis; slow, kinking, rolling, coiling, or twitching motions; and hypercontraction—have been isolated.

Complementation analysis has defined about 100 genes, called *unc* (for uncoordinated), responsible for these behaviors. Of course, an uncoordinated phenotype may result from alteration in any of the various structures needed for normal movement, and at least some of these mutations must result from alterations in the muscle or in the nervous system of the animal. In fact, the possibility offered by *C. elegans* of integrating different approaches has allowed the classification of most of the *unc* genes into those affecting muscle and those affecting the nervous system.

Using biochemical, microscopic, and genetic techniques, the structural gene for one of the myosin heavy chains has been identified (Epstein et al., 1975) and, more recently, cloned and sequenced (MacLeod et al., 1981). In *C. elegans* three myosin heavy chains can be identified biochemically. One of them, with molecular weight (MW) of 210,000, is present in both the pharynx and body wall muscles; one, 206,000 MW, is present only in the pharynx; and one, 210,000 MW, only in the body wall. The isolation of muscle proteins from one allele of *unc*-54 shows a new protein of 203,000 MW that behaves like a myosin heavy chain. The protein has been chemically analyzed and shown to be the result of an internal deletion of about 100 amino acids. This conclusion has been confirmed by the analysis of the cloned gene. In *unc*-54 mutants, movement is slow and body wall muscles show disruption of the sarcomere banding pattern with a reduced number of thick filament, whereas the muscles of the pharynx are normal and the animal can feed and survive. Thus, gene *unc*-54 specifies the myosin heavy chain that is present in the body wall muscle but absent from the pharynx muscle. Monoclonal antibodies identifying each of the three myosin heavy chains have been obtained and are being used for the study of the appearance of the different myosins during development.

The structural gene for paramyosin has also been tentatively identified, and mutations in it (*unc*-15) affect both pharnyx and body wall muscle, although the lack of paramyosin has a much less drastic effect on pharyngeal than on body wall muscle. Extragenic suppressors of *unc*-54 and *unc*-15 mutations have been sought in order to identify other genes involved in the specification of muscle structure—for example, to identify proteins that interact with myosin. Mutations identifying genes involved in sarcomere assembly and organization of myofilaments have been identified (reviewed by Zengel and Epstein, 1980).

Although the analysis of *unc* mutants in which the nervous system is affected is more difficult, at least two kinds of defects have already been recognized by means of optic and electronic microscopy in conjunction with our detailed anatomical and developmental knowledge about the nervous system of the wild type. In certain *unc* mutants, the wiring of the nervous system is altered such that illicit branching and incorrect turns and/or premature termination of processes occurs (Hedgecock, personal communication, 1981). In others, the phenotype can be correlated with a lack or excess of certain neurons and is due to alterations in the division pattern of the cellular precursors of these neurons (Sulston and Horvitz, 1981).

Genetic analysis, optic and electronic microscopy, and laser ablation experiments have elucidated some aspects of the sensory apparatus involved in the response to touch of *C. elegans*. The worm reacts to gentle mechanical stimuli by moving away. Six mechanoreceptor cells, also called touch cells or microtubule cells, are responsible for this sensitivity. Laser ablation of these cells completely eliminates the response without apparent deleterious effects. The lack of the response to touch of animals otherwise able to move has been used to isolate over 50 touch-insensitive mutants; these fall into 13 complementation groups called *mec* for mechanoreceptor. Some of the mutations affect various aspects of the ultrastructure of the microtubule cells. Examples include mutations in which the special microtubules (see below) are missing. In other cases, synapse formation is defective or the growth of the receptor process is stunted. These structural defects naturally cause impaired function of the cells. Other genes affect the development of the touch cells; when mutated, one gene causes the selective death of all microtubule cells; another alters the lineage pattern of the precursors of the touch cells, resulting in their complete absence (Chalfie, personal communication, 1981).

Electron-microscopic analysis of the touch cells shows that over 95% of the microtubules of these cells have 15 protofilaments instead of the 11 protofilaments characteristic of the cytoplasmic microtubules of the other cells. The 11 protofilament microtubules seem to be involved in such processes as mitosis, cell architecture, neuron outgrowth, and so forth, whereas the 15 protofilament microtubules seem specifically responsible for sensory transduction of mechanical stimuli. Eleven- and 15-protofilament microtubules are affected by different mutations and show different sensitivities to certain antimitotic drugs (colchicine, podophyllotoxin, etc.)

(Chalfie and Thomson, 1979). It is worth noting that microtubules in almost all other eukaryotes have 13 protofilaments and that the 11- and 15-protofilament arrangements may be shared by all nematodes, since these are observed in another nematode, *Panagrellus redivivus*. It is hoped that a thorough study of these differences between nematodes and other eukaryotes may serve as the basis to improve the selectivity of antihelmintic drugs (M. Chalfie, personal communication, 1981).

Other aspects of *C. elegans* behavior, such as chemotaxis and thermal adaptation, as well as the role of neurotransmitters and their receptors, have been analyzed genetically, but it is not possible to review the findings here.

Developmental Mutants

With regard to mutations affecting postembryonic development, those affecting the lineage of certain nerve cells have been mentioned briefly. Another particularly interesting group are those affecting the formation of the vulva. Their isolation has been simplified because a serious vulval defect prevents egg laying; in such cases, the eggs hatch inside the female, transforming it into a "bag of worms" that floats in solutions of appropriate density, whereas wild-type hermaphrodites sink.

As shown in Fig. 5, the *C. elegans* hermaphrodite vulva is formed by 22 cells deriving from the posterior daughters of P5, P6, and P7. (P1 through P12 are ventral hypodermal cells that divide during postembryonic development to give neurons, more hypodermal cells, and the vulva in the hermaphrodite) (Sulston and Horvitz, 1977). The presence of a developing hermaphrodite gonad triggers the special pattern of division of P5, 6, and 7, since after laser ablation of the gonad primordium, P5, 6, and 7 produce only 6 of the 22 cells of the vulva (Kimble et al., 1979; Sulston and White, 1981).

More than 80 mutations affecting vulval development have been isolated. They identify 15 genes and fall into two major classes: vulvaless *(vul)* and multivulval *(muv)*. In vulvaless mutants, P5p, P6p, P7p, or some of their descendants either fail to divide or do so in an incorrect pattern. In multivulval mutants P3p through P8p, all divide at least three times, forming up to five abnormal vulvas. The study of the interaction (epistatic effects) between different *vul* and *muv* mutations can help explain whether the mutational defect is due to increased or decreased responsiveness of the P cells to the gonadal signal or whether the signal itself is lacking or too strong (Horvitz, personal communication, 1981).

A special feature of the life cycle of nematodes, including *C. elegans*, that has great survival value is the formation of dauer larva. In the absence of food, a second-stage larva will molt and produce a special form, the dauer larva, that can withstand adverse conditions for up to 60 days and is distinguishable morphologically and behaviorally from normal larvae. When conditions improve (e.g., food is again available), the dauer larva begins a period of recovery, and after molting it reenters the normal developmental cycle.

Riddle (1977) has isolated two types of dauer mutants: dauer-defectives, which are unable to become dauers, and dauer-constitutive, which become dauers even

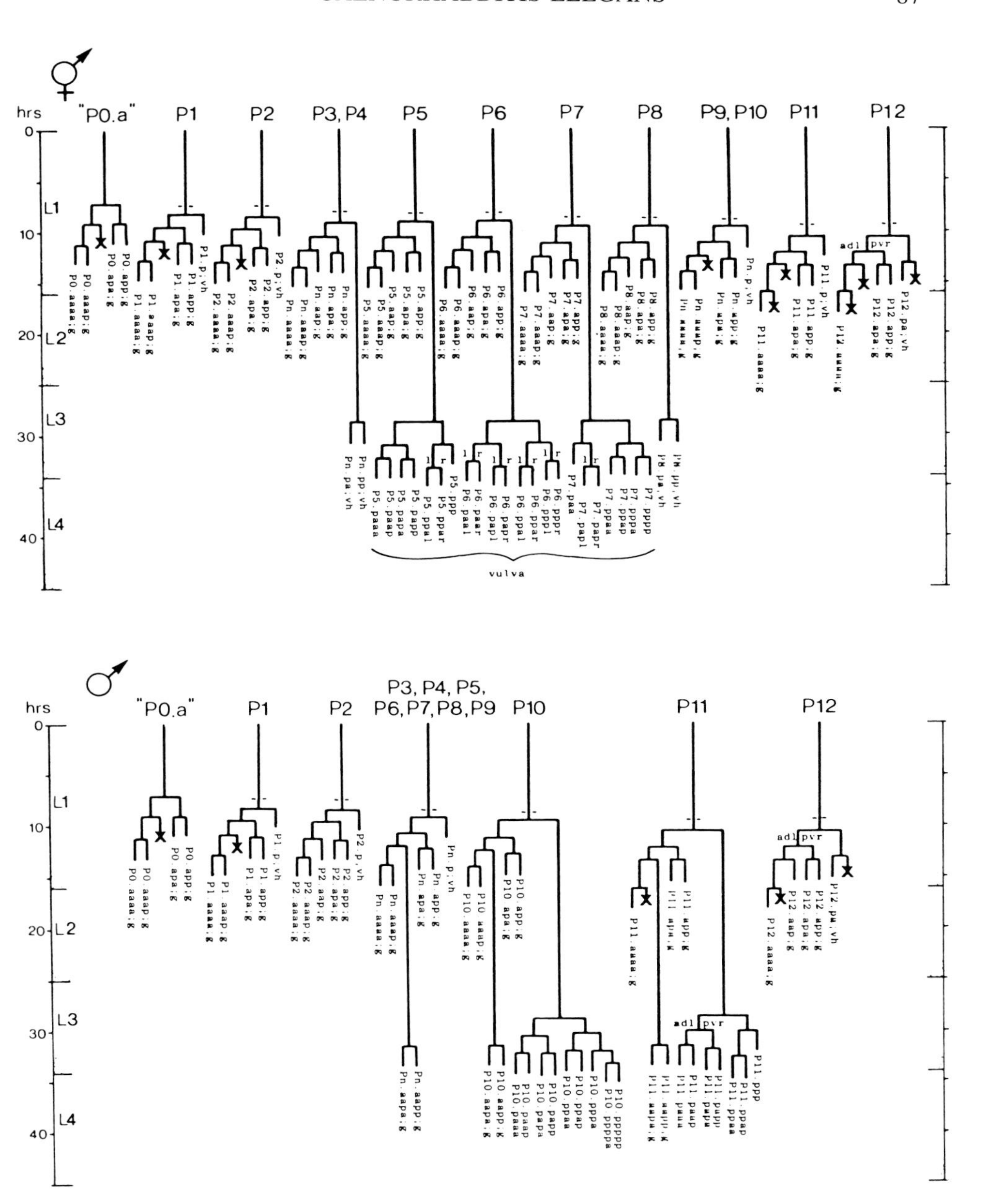

FIG. 5. Postembryonic divisions of P blast cells after their migration into the ventral cord. g, neuron or glial cell; vh, ventral hypodermal cell. The 16 extra cells produced in the male by P10.p and P11.p become male-specific nuclei in the region of the preanal ganglion. The postembryonic blast cells P shown here are different from the germ line precursors P1–P4 present in the embryo (Fig. 3). (From Sulston and Horvitz, 1977, with permission.)

under normal environmental conditions. Since this latter class would result in sterility, they are useful as temperature-sensitives. Some of the dauer-defective mutants are also defective in chemotaxis and have morphological defects in the amphids, which suggests that these are the receptors for the environmental signal that triggers the dauer pathway. By determining the phenotype of strains containing both a dauer-constitutive and a dauer-defective mutation, Riddle has constructed a developmental pathway for dauer formation (Fig. 6), in much the same way that biochemical geneticists have constructed metabolic pathways. Riddle has explained the rationale behind the pathway in a very clear way (Riddle, 1980):

> The dauer-defective mutants are blocked in the "natural" pathway of dauer formation. The dauer constitutives, on the other hand, generate a false internal signal which causes the mutant to form a dauer even in the absence of the natural signal (which accompanies starvation). If the pathway is blocked before the false signal, the double mutant will be dauer defective. If the false signal is generated after the block, the double mutant will be dauer constitutive.

As one might expect, mutants with altered amphids are blocked early in the pathway. Studies in progress have tentatively identified the substance(s) that signal the worm to enter the dauer pathway and/or inhibit the reentry into the normal developmental cycle (J. W. Golden and D. L. Riddle, personal communication, 1981). The importance of this research for nematode control is evident.

Embryonic development, sex determination, gonadogenesis, spermatogenesis, and fertilization are other aspects of development that have also been analyzed genetically with very interesting results.

Biochemical Mutants

Under this group I briefly list mutants that define genes whose products are known. Four genes (*flu*-1, -2, -3, and -4) affecting tryptophan catabolism have been identified, thanks to the change in the intensity and color of the fluorescence of granules present in the cytoplasm of intestinal cells. *Flu*-1 seems to affect kinurenine hydroxylase and *flu*-2, kinureninase (Siddiqui and Von Ehrenstein, 1980). *Ace*-1

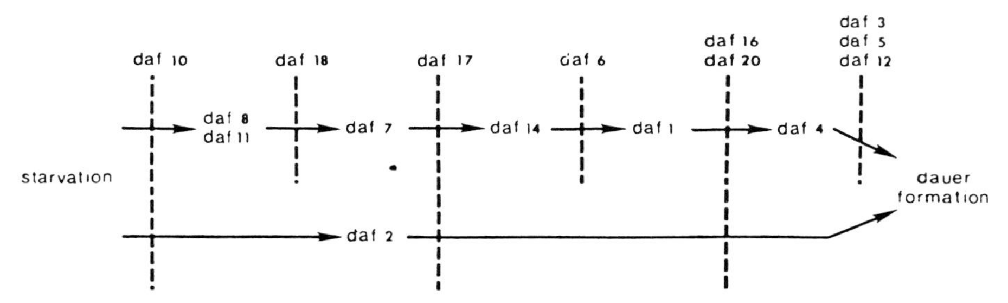

FIG. 6. A partial genetic pathway of dauer larva formation based on epistatic relationships between dauer-constitutive and dauer-defective mutations. Gene names of dauer constitutives are given in the pathway itself to represent the points where false signals can be initiated. Dauer-defective genes block the pathway at the positions shown by dashed lines. (From Riddle, 1980, with permission.)

and *Ace*-2 are probably the structural genes for the two acetylcholinesterases present in the worm (J. Culotti et al., personal communication, 1981). *Cha*-1 is the putative structural gene for choline acetyl-transferase (Russel et al., 1977). The mutant *nuc*-1 lacks an endonuclease responsible for digestion of the DNA in the intestinal lumen (Sulston, 1976). Six mutants having 10^{-3} of the β-glucuronidase activity of wild types have been isolated. They fall into one complementation group that probably identifies the structural gene for this enzyme. The mutations map on the right arm of linkage group 1 (Bazzicalupo, unpublished).

STUDIES OF THE DNA OF *C. elegans*

One of the advantages offered by *C. elegans* is the small size of its genome. The DNA content of the haploid genome is about 8×10^7 base pairs; this is roughly 20 times the *E. coli* genome or half the genome of *Drosophila* and is the smallest genome reported for a multicellular animal. The number of essential genes is estimated at between 2,000 and 4,000 (Brenner, 1974; Herman, 1978). This gives a figure of 20 to 40 $\times$ 10^3 base pairs per essential gene, which is close to that for *Drosophila* (~30,000). About 83% of the DNA consists of unique sequences; the remaining 17% of repetitive sequences. The repetitive DNA is kinetically heterogeneous, with repetitiveness ranging more or less continuously from 1 to 2,000 and without clearly predominant classes (Emmons et al., 1980). Ribosomal DNA is present in about 50 copies per haploid genome, has a guanine plus cytosine (GC) content of 51%, and bands as a satellite because the bulk DNA is 36% GC. The rDNA genes are clustered, and the repeating units, which show no sequence or length heterogeneity, are 6,800 base pairs long and contain the 18S, 5S, and 28S genes with a nontranscribed region of not more than 1,000 base pairs (Files and Hirsh, 1981).

Using cloned fragments of *C. elegans* DNA, Emmons et al. (1979) compared somatic DNA and germ line (sperm) DNA to test for the occurrence of significant rearrangements of the DNA during somatic differentiation. At the beginning of the century, Boveri observed the phenomenon of chromatin diminution in the parasitic nematode *Ascaris*: during the early cleavages of the fertilized zygote, at each division, one cell, the precursor of the germ line, keeps all its chromosomal material, while its sister, which will form somatic tissues, loses some. The only cells that retain the complete genome are, naturally, the germ cells. In *Ascaris* the loss of DNA in somatic tissues amounts to ~30%. In contrast, Emmons et al. (1979) found that the DNA in *C. elegans* does not undergo extensive rearrangements during development, although their experiment does not rule out the occurrence of minor rearrangements involving less than a few percent of the DNA.

With regard to the study of the DNA of particular genes, the cloning and sequencing of the gene for one of the myosin heavy chains (MacLeod et al., 1981) has already been mentioned. Cloning of rDNA (Files and Hirsh, 1981) and tRNA (Cortese et al., 1978) has also been accomplished. Sequences specifying proteins that are highly conserved in evolution are being identified and isolated, taking advantage of DNA probes prepared from other organisms. I mention here only the isolation and cloning of actin genes. Files and Hirsh (personal communication) have used a probe for actin sequences made from the slime mold *Dyctostelium discoideum*. They have shown that there are four actin genes in *C. elegans*, and they have cloned and started to sequence them. Three of the genes are clustered within a 12,000-base-pair region of DNA, and the orientation of transcription has been determined. Like the myosin gene, they have introns; the intron-exon junctions are conserved and are identical to those of the myosin gene. The introns are not found in the same position in the three genes.

Using a somewhat unconventional approach, Hirsh and his collaborators (personal communication) have mapped the cluster of the three actin genes on linkage group 5. Two cross-fertile strains of *C. elegans*, variety Bristol and variety Bergerac, are commonly used in the laboratory. In the Bergerac strain, but not in the Bristol, there is an insertion 1,700 base pairs long next to one of the actin genes in the cluster. This insertion causes a difference between the two strains in the pattern of DNA fragments hybridizing on a Southern filter to an actin probe. Twelve sets of hybrids between the two strains have been constructed and DNA prepared from each of them. In each set, one chromosome is held homozygous for either Bristol or Bergerac, whereas the other chromosomes are allowed to become heterozygous. Southern hybridization of an actin probe to these DNAs allows the assignment of the cluster of actin genes to the chromosome that, when held homozygous, will show only either the Bristol or the Bergerac pattern of hybridization.

Several other laboratories are exploiting recombinant DNA technology to isolate and study a variety of interesting genes. The addition of this powerful new approach to the study of *C. elegans* will undoubtedly help explain its biology at the molecular level.

CONCLUSION

The opportunities that *C. elegans* offers of integrating various experimental approaches makes this nematode a very good model system for the study of the biology of multicellular eukaryotes at the molecular level. This organism has so far been studied mainly by researchers interested in the most basic biological problems, and the work and the results obtained naturally reflect their perspective. Although the information about the biology of this one species is very useful in studying other nematodes, including the parasitic species, it would be extremely worthwhile if *C. elegans* were also made the object of research aimed more specifically at understanding those aspects of life cycle, physiology, and structure of nematodes that may be relevant in the parasitic mode of existence. Introducing this organism to parasitologists is one step in that direction.

REFERENCES

1. Albertson, D. G., and Thomson, J. N. (1976): The pharynx of *Caenorhabditis elegans*. *Phil. Trans. R. Soc. London, Ser. B.*, 275:299–325.
2. Brenner, S. (1974): The genetics of *Caenorhabditis elegans*. *Genetics*, 77:71–94.

3. Chalfie, M., and Thomson, J. N. (1979): Organization of neuronal microtubules in the nematode *Caenorhabditis elegans. J. Cell. Biol.*, 82:278–289.
4. Cortese, R., Melton, D., Tranquilla, T., and Smith, J. D. (1978): Cloning of nematode tRNA genes and their expression in frog oocyte. *Nucleic Acids Res.*, 12:4593–4611.
5. Croll, N. A., editor (1976): *The Organization of Nematodes.* Academic Press, New York.
6. Deppe, U., Shierenberg, E., Cole, T., Krieg, C., Schmitt, D., Yoder, B., and Von Ehrenstein, G. (1978): Cell lineages of the embryo of the nematode *Caenorhabditis elegans. Proc. Natl. Acad. Sci. USA*, 75:327–342.
7. Dougherty, E. C., Hansen, E. L., Nicholas, W. L., Mollett, J. A., and Yarwood, E. A. (1959): Axenic cultivation of *Caenorhabditis briggsiae* with supplemented and unsupplemented chemically defined media. *Ann. N.Y. Acad. Sci.*, 77:176–217.
8. Emmons, S. W., Klass, M. R., and Hirsh, D. I. (1979): Analysis of the constancy of DNA sequences during development and evolution of the nematode *Caenorhabditis elegans. Proc. Natl. Acad. Sci. USA*, 76:1333–1337.
9. Emmons, S. W., Rosenzweig, B., and Hirsh, D. I. (1980): The arrangement of repeated sequences of the nematode *Caenorhabditis elegans. J. Mol. Biol.*, 144:481–500.
10. Epstein, H. F., Waterstone, R. H., and Brenner, S. (1975): A mutant affecting the heavy chain of myosin in *Caenorhabditis elegans. J. Mol. Biol.*, 90:291–300.
11. Files, J. G., and Hirsh, D. I. (1981): The ribosomal DNA of *Caenorhabditis elegans. J. Mol. Biol.*, 149:223–240.
12. Harris, J. E., and Crofton, H. D. (1957): Structure and function in the nematodes: internal pressure and cuticular structure in *Ascaris. J. Exp. Biol.*, 34:116–130.
13. Herman, R. K. (1978): Crossover suppressors and balanced recessive lethals in *Caenorhabditis elegans. Genetics*, 88:49–65.
14. Herman, R. K., and Horvitz, R. H. (1980): Genetic analysis of *Caenorhabditis elegans.* In: *Nematodes as Biological Models*, Vol. 1, edited by B. M. Zuckerman, pp. 227–261. Academic Press, New York.
15. Hodgkin, J., Horvitz, R. H., and Brenner, S. (1979): Nondisjunction mutants of the nematode *Caenorhabditis elegans. Genetics*, 91:67–94.
16. Horvitz, R. H., and Sulston, J. E. (1980): Isolation and genetic characterization of cell lineage mutants of the nematode *Caenorhabditis elegans. Genetics*, 96:435–454.
17. Kimble, J., and Hirsh, D. I. (1979): The post-embryonic cell lineages of the hermaphrodites and male gonads in *Caenorhabditis elegans. Dev. Biol.*, 70:396–417.
18. Kimble, J., Sulston, J. E., and White, J. G. (1979): Regulative development in the post-embryonic lineages of *Caenorhabditis elegans.* In: *Stem Cells, Cell Lineages, and Cell Determination*, edited by N. leDouarin. Elsevier, North-Holland, Amsterdam.
19. Kunhelt, W., Walker, N., Butcher, J. W., and Laughlin, C. (1976): *Soil Biology with Special Reference to the Animal Kindgom.* Michigan State University Press, East Lansing.
20. Lee, D. L. (1965): *The Physiology of Nematodes.* Oliver and Boyd, Edinburgh.
21. McLeod, A. R., Karn, J., and Brenner, S. (1981): Molecular analysis of the *unc*-54 myosin heavy-chain of *Caenorhabditis elegans. Nature*, 291:386–390.
22. Riddle, D. L. (1977): A genetic pathway for dauer larva formation in the nematode *Caenorhabditis elegans. Genetics*, 86:S51–S52.
23. Riddle, D. L. (1980): Developmental genetics of *Caenorhabditis elegans.* In: *Nematodes as Biological Models*, Vol. 1, edited by B. M. Zuckerman, pp. 263–283. Academic Press, New York.
24. Russell, R. L., Johnson, C. D., Sherer, S., and Zwass, M. S. (1977): Mutants of acetyl-choline metabolism in the nematode *Caenorhabditis elegans.* In: *Molecular Approaches to Eukaroytic Genetic Systems. Symposia on Molecular Cellular Biology*, Vol. 8, edited by G. Wilcox, J. Abelson, and C. F. Fox, pp. 359–371. Academic Press, New York.
25. Siddiqui, S. S., and Von Ehrenstein, C. (1980): Biochemical genetics of *Caenorhabditis elegans.* In: *Nematodes as Biological Models*, Vol. 1, edited by B. M. Zuckerman, pp. 285–304. Academic Press, New York.
26. Sulston, J. E. (1976): Post-embryonic development in the ventral cord of *Caenorhabditis elegans. Phil. Trans. R. Soc. London, Ser. B*, 275:287–298.
27. Sulston, J. E., and Brenner, S. (1974): The DNA of *Caenorhabditis elegans.* Genetics, 77:95–104.
28. Sulston, J. E., and Horvitz, R. H. (1977): Post-embryonic cell lineages of the nematode *Caenorhabditis elegans. Dev. Biol.*, 56:110–156.

29. Sulston, J. E., and Horvitz, R. H. (1981): Abnormal cell lineages in mutants of the nematode *Caenorhabditis elegans*. *Dev. Biol. (in press).*
30. Sulston, J. E., and White, J. G. (1981): Regulation and cell autonomy during post-embryonic development in *Caenorhabditis elegans*. *Dev. Biol. (in press).*
31. Ward, S., Thomson, J. N., White, J. G., and Brenner, S. (1975): Electron microscopical reconstruction of the anterior sensory anatomy of the nematode *Caenorhabditis elegans*. *J. Comp. Neurol.*, 160:313–338.
32. White, J. G., Southgate, E., Thomson, J. N., and Brenner, S. (1976): The structure of the ventral nerve cord of *Caenorhabditis elegans*. *Phil. Trans. R. Soc. Lond., Ser. B*, 275:327–342.
33. Zengel, J. M., and Epstein, H. F. (1980): Muscle development in *Caenorhabditis elegans:* a molecular genetic approach. In: *Nematodes as Biological Models*, Vol. 1, edited by B. M. Zuckerman, pp. 73–126. Academic Press, New York.
34. Zuckerman, B. M., editor (1980): *Nematodes as Biological Models*. Academic Press, New York.

Molecular Biology of Parasites, edited by
J. Guardiola, L. Luzzatto, and W. Trager.
Raven Press, New York © 1983.

Metazoan Parasites: The Schistosomes. An Introductory Guide to the Literature

Donato Cioli

Laboratory of Cell Biology, National Research Council, 00196 Rome, Italy

It has been estimated that about 200 million people living in tropical and subtropical countries are currently affected by schistosomiasis (bilharziasis). The disease is caused by trematodes (flatworms) belonging to different species of the genus *Schistosoma* (73). *S. mansoni* and *S. japonicum* cause intestinal schistosomiasis, while *S. hematobium* causes the urinary form of the disease (39). The life cycle of the parasite involves two hosts, i.e., a mammalian (final) host and a snail (intermediate) host. The transition between the two hosts is effected via a free-swimming larva that in the case of the form derived from the mammal is called a miracidium, while in the case of the form emerging from the snail is called a cercaria. Both types of free-swimming larvae are short-lived (about 24 hr) (43) and achieve infection of the respective host by direct penetration of exposed body surfaces (skin, tegument).

LIFE CYCLE

Cercariae

Cercariae emerge from infected snails following the rupture of special formations induced by the parasite and called sporocysts. Cercarial emergence is stimulated by light (1), so that under field conditions, maximal shedding of cercariae is usually observed in the middle part of the day. Problems concerning the relative contribution of the snail host and of the parasite in various morphological and functional phenomena can be now approached by the use of tissue culture techniques that are available for snail cells (5,34).

Cercariae consist of a head and of a forked tail, which assists them in the swimming motions. Swimming is usually performed tail-up towards the surface of water. Upon reaching the surface, movements stop, and the cercaria begins to sink slowly. After a while, swimming is resumed, so that the general behavior consists of short periods of activity alternating with pauses. Cercariae are sensitive to thermal stimuli, being attracted by bodies with a temperature higher than the water temperature (22). They are also sensitive to chemical stimuli, as can be easily shown *in vitro* by observing penetration of cercariae into agar layers containing various skin lipid components (32,45,60).

Upon contact with a suitable mammalian host, cercariae attach to its skin and initiate a series of remarkably efficient activities that result in their penetration in a matter of about 5 min (12). These activities consist of mechanical movements and of chemical reactions. The latter are due to the secretions contained in a series of glands in the cercarial head. An impressive collection of enzymes is present in the secretion of these glands, which are emptied in a given sequence during penetration (66,68). Once the head has gone into the stratum corneum, the tail is detached, and it is lost in the outside water (37). A large amount of energy is spent for penetration, and this is reflected in a dramatic drop of the glycogen content of cercariae (13).

Schistosomula

The head of the penetrated cercaria is called a schistosomulum. Profound structural and functional changes occur in the transition from the water-swimming cercaria to the tissue-dwelling schistosomulum (65). Osmotic changes are particularly impressive, since schistosomula that have recently penetrated a host die by osmotic lysis if placed back into water (20). The main structural change occurring in the cercaria-schistosomulum transition is represented by the loss of a thick surface coat of mucopolysaccharides (glycocalyx), which probably renders the cercarial surface impermeable (19). Another important structural change occurs in the tegument of schistosomula within the first few hours after penetration. This consists in the formation of a double outer membrane, i.e., in the appearance of a double lipid bilayer on the surface of schistosomula (35). This surface structure (which has an heptalaminate appearance in electron-microscopic sections) is a characteristic common to all blood-dwelling trematodes (48).

After traversing the epidermis and the dermis, schistosomula penetrate into blood vessels and are carried into the general circulation (49). They reach the right heart and then the lungs, where they can be found in maximal numbers about 5 days after infection. The subsequent route of migration from the lungs to the liver has been the object of some uncertainty until quite recently (40,74). It now seems well established (50) that such migration is via arterial blood returning to the left heart from the lungs and then into the general circulation via the aorta. Those parasites that enter a mesenteric artery reach the portal circulation and stop in the liver; those that enter different areas will recirculate through the heart and lungs, possibly several times, until they reach their proper location in the portal tissue. Metabolic changes accompany the migration of schistosomula through various tissues (23,42).

Adult Worms

A phase of rapid growth and differentiation occurs in the liver. Male and female worms differentiate quite markedly, and pairing of the two sexes occurs after the 3rd week from infection. The mechanisms involved in worm pairing are only partially understood, but they may include some type of chemical attraction between the two sexes (38). Paired worms now begin the last phase of migration, i.e., they progress (against the blood flow) from the liver to the small venules of the mesenteric circulation. The presence of male worms is necessary for the final steps of development of the female reproductive apparatus (29), but the mechanisms and the possible chemical signals involved in this stimulation of maturation are still under investigation (2,58).

If, after reaching their final location in the mesenteric venules, schistosomes are subjected to unfavorable, stressful conditions [as after administration of certain drugs (57) or anesthetics (28) to the host], they abandon the mesenteric veins to be passively swept by the bloodstream to the portal vein and to the liver. However, if this condition is short-lived, schistosomes will soon migrate back to their mesenteric location.

The mechanisms controlling this migration of adult worms are almost completely unknown (3), but the requirement for a site as close as possible to the intestinal wall is of obvious teleological significance for the subsequent fate of deposited eggs.

Eggs

Each pair of adult schistosomes is capable of producing, on the average, one egg every 3 to 5 min (69). Their shape, including a characteristic spine, is species-specific. The spine probably helps the egg in sticking to the internal wall of blood vessels, while various secretions from the porous shell attack the various layers of the vein wall. Histolytic activity continues beyond the blood vessel into the intestinal wall, until the egg breaks into the lumen of the intestine, mixes with the feces, and eventually reaches the external environment. In *S. hematobium* infections the final location of adult worms is in the veins of the vescical plexus, and eggs are voided into the urinary bladder. Eggs that happen to reach a body of fresh water will hatch and produce a free-swimming miracidium. Miracidia are capable of infecting the specific aquatic snail (4); they multiply asexually inside the molluscan host, and finally emerge as infective cercariae.

Only a portion of the eggs deposited by schistosomes actually reaches the external environment, and a large number of eggs remain in the host body, either trapped in the intestinal tissues or swept to the liver by the bloodstream. Eggs that accumulate in the liver will elicit a characteristic granuloma, which is the result of a delayed hypersensitivity type of reaction (33,71). One of the features of advanced disease is the progressive impairment of portal blood circulation because of the granulomatous reaction, portal hypertension, and, eventually, a collection of ascitic fluid in the abdominal cavity (16). Another consequence of portal hypertension is the opening of shunts between the portal and the caval circulation, so that the blood coming from the intestine is discharged into the vena cava and returns to the heart without passing through the liver.

IMMUNE RESPONSES

Adult schistosomes are capable of an extraordinarily long life (up to several decades) in the bloodstream of their mammalian host (6,72). Their long survival is particularly impressive, since it occurs in spite of a large variety of immune responses that are elicited in the host (61,63). Thus, antibodies in serum can reach extremely high levels (10), and cells can be shown to be specifically sensitized using a number of *in vitro* tests (54). There is also abundant evidence to show that hosts infected with a small number of schistosomes are largely resistant to subsequent reinfection with the same parasite (54). Under these conditions, it is the newly acquired schistosomula of the challenge infection that are killed in the host, whereas the established adult schistosomes of the primary infection survive, completely unaffected. This situation has been termed "concomitant immunity" (62), and a considerable amount of research has been carried out, especially in the mouse, to elucidate the mechanisms of the phenomenon and, hopefully, to isolate the parasite antigens that may be responsible for resistance to reinfection. It should be stressed, however, that attempts to produce resistance using extracts of all types, instead of a live infection, have thus far completely failed or given marginal and poorly reproducible results (61).

Infection with worms of a single sex (i.e., in the absence of egg production) does not induce significant resistance to reinfection, at least in the mouse model (7,27). Passive transfer of resistance from an infected host to a naive animal, although occasionally successful, has generally failed when using mouse serum (25). The recent elucidation of the route of postlung migration of schistosomes (50) has generated some doubts as to whether mouse resistance to reinfection is really an immune phenomenon. It could be that chronically infected hosts have developed a collateral circulation that is sufficient to shunt most of the blood from the portal to the caval circulation. Under these circumstances, developing schistosomula of the challenge infection would never reach the liver and so accomplish the crucial phases of growth, differentiation, and pairing that normally take place solely in the liver (75).

An extremely promising new experimental model involves infecting mice with highly irradiated cercariae (8,51). These cercariae are capable of penetrating and of surviving for a few days, but following high levels of irradiation, they cannot complete migration and maturation and do not deposit eggs. In this model, most pathological alterations are absent, yet a good level of resistance can be achieved and can be passively transferred to naive hosts (25).

Numerous *in vitro* systems have been developed to take advantage of the possibility of artificially transforming cercariae into schistosomula and of subsequently growing them in the test tube (21). Worms will differentiate and pair *in vitro*, but no deposition of viable eggs has been achieved under *in vitro* conditions (18). Newly transformed schistosomula can be killed *in vitro* by antibody plus complement, by antibody and cells, or by mixtures of various components (54). One of the most interesting findings that has emerged from these *in vitro* studies is that eosinophils are often the cells that appear to perform the actual killing of schistosomula (14,15). Somewhat parallel to the *in vivo* situation, only newly transformed schistosomula can be killed in these *in vitro* systems, while schistosomula cultured for 1 or 2 days in normal medium become generally refractory to the same *in vitro* mechanisms (21). The relevance of *in vitro* studies to the *in vivo* situation is hard to evaluate, but there is one report (44) indicating that eosinophils may also be important in the *in vivo* resistance to a challenge infection.

MECHANISMS OF SURVIVAL

An attractive hypothesis was proposed in 1969 by Smithers et al. (64) to explain both the long survival of adult schistosomes and the phenomenon of concomitant immunity. According to this hypothesis, the surface of adult schistosomes becomes coated with host antigenic molecules, which would mask parasite antigens and thus render the worms unsusceptible to host immune responses. Young developing schistosomula, which do not have time to adopt this "disguise," would be vulnerable to host attack, a possible explanation for their death under conditions of concomitant immunity.

Experimental support for this hypothesis was derived from results showing that adult schistosomes survive when transplanted from a mouse to a monkey but die when transplanted from a mouse to a monkey that has been preimmunized with normal mouse antigens (64). The same phenomenon occurs with different host systems (17), and it could be shown that the host antigens were passively acquired by the parasites after a few days in a given host species (17,64). Blood group substances (30) and, more recently, products of the major histocompatibility complex (59) have been detected with various methods on the surface of schistosomes. In addition, it has been shown with a variety of different techniques that newly penetrated schistosomula expose parasite antigens and lack host antigens, whereas the opposite is true for 4-day-old worms recovered from the lungs of mice (31,47).

While the existence of hostlike molecules on the surface of schistosomes is a well established fact, the actual involvement of these molecules in protecting the parasite from the host immune attack is still the object of an active debate. For instance, it has been shown that schistosomula grown for a couple of days *in vitro* in a chemically defined medium become insensitive to killing by antibody and complement even in the absence of any host antigenic material (24). Similarly, that the loss of susceptibility to immune attack cannot be attributed solely to host antigen disguise is shown by the fact that trinitrophenyl (TNP) groups attached to the surface of young (skin) schistosomula render these organisms susceptible to killing by anti-TNP antibody, while older (lung) worms covered with the same hapten are unsusceptible (52). In addition, there are certain discrepancies in the time at which host antigens are acquired as compared to the time at which protection is acquired by the parasite. Host antigens appear on the worm surface in a matter of hours after penetration, and by the 4th day, they seem to mask schistosome antigens completely (47), while it has been shown that immune killing of the worms may occur at later stages, possibly up to 12 days after penetration (9,27).

Another possible mechanism of survival could be represented by the fact that schistosomes undergo rapid membrane turnover (41) and might continually replace those surface structures that have been attacked and damaged by antibody or by cells (53). When released in the host circulation, these antigens might contribute to a decreased efficiency of the immune response, possibly through the formation of immune complexes (55). *In vitro* studies have provided evidence for shedding (sloughing) of surface antigens (11). In certain experimental systems, some surface antigens may be lost by young schistosomula and not be replaced in older worms (56).

Recent investigations have focused on the detailed molecular organization of the worm surface during schistosome development. Freeze-fracture studies have revealed that the intramembranous particles (IMP) of the two lipid bilayers undergo significant changes in number and distribution during parasite maturation (36,70). In particular, the IMP associated with the external face of the outer bilayer reach their peak density on day 4 and appear unusually large and evenly distributed. It has been suggested that these IMP may represent complexes of parasite and host antigens (46). Although it may be premature to advance a functional hypotheses based on the available data, the freeze-fracture technique seems a very promising approach for a detailed study of the worm surface.

In conclusion, no single mechanism is completely satisfactory in accounting for the long survival of schistosomes. Host antigens, surface turnover, and membrane reorganization may all play a role in protecting the schistosome from the host attack, or alternatively, totally different mechanisms may be operative and may be completely beyond our present understanding.

ACKNOWLEDGMENTS

Research conducted in the author's laboratory is currently receiving financial support from UNDP/World Bank/WHO Special Programme for Research and Training in Tropical Diseases.

REFERENCES

1. Asch, H. L. (1972): Rhythmic emergence of *S. mansoni* cercariae from *B. glabrata*: control by illumination. *Exp. Parasitol.*, 31:350–355.
2. Atkinson, K. H., and Atkinson, B. G. (1980): Biochemical basis for the continuous copulation of female *S. mansoni*. *Nature*, 283:478–479.
3. Awwad, M., and Bell, D. R. (1978): Fecal extract attracts copulating schistosomes. *Ann. Trop. Med. Parasitol.*, 72:389–390.
4. Basch, P. F. (1976): Intermediate host specificity in *S. mansoni*. *Exp. Parasitol.*, 39:150–169.
5. Basch, P. F., and DiConza, J. J. (1977): In vitro development of *S. mansoni* cercariae. *J. Parasitol.*, 63:245–249.
6. Berberian, D. A., Paquin, H. O., Jr., and Fantauzzi, A. (1953): Longevity of *S. hematobium* and *S. mansoni*: Observations based on a case. *J. Parasitol.*, 39:517–519.
7. Bickle, Q., Bain, J., McGregor, A., and Doenhoff, M. (1979): Factors affecting the acquisition of resistance against *S. mansoni* in the mouse. III. The failure of primary infections with cercariae of one sex to induce resistance to reinfection. *Trans. R. Soc. Trop. Med. Hyg.*, 73:37–41.
8. Bickle, Q., Taylor, M. G., Doenhoff, M., and Nelson, G. S. (1979): Immunization of mice with gamma-irradiated intramuscularly injected schistosomula of *S. mansoni*. *Parasitology*, 79:209–222.
9. Blum, K., and Cioli, D. (1981): *S. mansoni*: Age-dependent susceptibility to immune elimination of schistosomula artificially introduced into preinfected mice. *Par. Immunol.*, 3:13–24.

10. Bout, D., Rousseaux, R., Carlier, Y., and Capron, A. (1980): Kinetics of classes and sub-classes of total immunoglobulins and specific antibodies to *S. mansoni* during murine infection. *Parasitology*, 80:247–256.
11. Brink, L. H., Krueger, K. L., and Harris, C. (1980): Stage-specific antigens of S. mansoni. In: *The Host-Invader Interplay*, edited by H. Van den Bossche, pp. 393–404. Elsevier/North-Holland, Amsterdam.
12. Bruce, J. I., Pezzlo, F., McCarty, J. E., and Yajima, Y. (1970): Migration of *S. mansoni* through mouse tissue. Ultrastructure of host tissue and integument of migrating larva following cercarial penetration. *Am. J. Trop. Med. Hyg.*, 19:959–981.
13. Bruce, J. I., Weiss, E., Stirewalt, M. A., and Lincicome, D. R. (1969): *S. mansoni*: Glycogen content and utilization of glucose, pyruvate, glutamate, and citric acid cycle intermediates by cercariae and schistosomules. *Exp. Parasitol.*, 26:29–40.
14. Butterworth, A. E. (1977): The eosinophil and its role in immunity to helminth infection. *Curr. Top. Microbiol. Immunol.*, 77:127–168.
15. Butterworth, A. E., Sturrock, R. F., Houba, V., Mahmoud, A. A. F., Sher, A., and Rees, P. H. (1975): Eosinophils as mediators of antibody-dependent damage to schistosomula. *Nature*, 256:727–729.
16. Cheever, A. W. (1965): A comparative study of *S. mansoni* infections in mice, gerbils, multimammate rats and hamsters. *Am. J. Trop. Med. Hyg.*, 14:211–226.
17. Cioli, D. (1976): *S. mansoni*: A comparison of mouse and rat worms with respect to host antigens detected by the technique of transfer into hamsters. *Int. J. Parasitol.*, 6:355–362.
18. Clegg, J. A. (1965): In vitro cultivation of *S. mansoni*. *Exp. Parasitol.*, 16:133–147.
19. Clegg, J. A. (1972): The schistosome surface in relation to parasitism. In: *Functional Aspects of Parasite Surfaces*, edited by A. E. R. Taylor, pp. 23–40. Blackwell, London.
20. Clegg, J. A., and Smithers, S. R. (1968): Death of schistosome cercariae during penetration of the skin. II. Penetration of mammalian skin by *S. mansoni*. *Parasitology*, 58:111–128.
21. Clegg, J. A., and Smithers, S. R. (1972): The effects of immune Rhesus monkey serum on schistosomula of *S. mansoni* during cultivation in vitro. *Int. J. Parasitol.*, 2:79–98.
22. Cohen, L. M., Neimark, H., and Eveland, L. K. (1980): *S. mansoni*: Response of cercariae to a thermal gradient. *J. Parasitol.*, 66:362–364.
23. Coles, G. C. (1973): The metabolism of schistosomes: A review. *Int. J. Biochem.*, 4:319–337.
24. Dean, D. A. (1977): Decreased binding of cytotoxic antibody by developing *S. mansoni*: Evidence for a surface change independent of host antigen adsorption and membrane turnover. *J. Parasitol.*, 63:418–426.
25. Dean, D. A., Bukowski, M. A., and Clark, S. S. (1981): Attempts to transfer the resistance of *S. mansoni*-infected and irradiated cercaria-immunized mice by means of parabiosis. *Am. J. Trop. Med. Hyg.*, 30:113–120.
26. Dean, D. A., Cioli, D., and Bukowski, M. A. (1981): Resistance induced by normal and irradiated *S. mansoni*. Ability of various worm stages to serve as inducers and targets in mice. *Am. J. Trop. Med. Hyg.*, 30:1026–1032.
27. Dean, D. A., Minard, P., Stirewalt, M. A., Vannier, W. E., and Murrell, K. D. (1978): Resistance of mice to secondary infection with *S. mansoni*. I. Comparison of bisexual and unisexual infections. *Am. J. Trop. Med. Hyg.*, 27:951–956.
28. Dickerson, G. (1965): Effect of anaesthetics on mature infections of *S. mansoni* in the white mouse. *Nature*, 206:953–954.
29. Erasmus, D. A. (1973): A comparative study of the reproductive system of mature, immature and unisexual female *S. mansoni*. *Parasitology*, 67:165–183.
30. Goldring, O. L., Clegg, J. A., Smithers, S. R., and Terry, R. J. (1976): Acquisition of human blood group antigens by *S. mansoni*. *Clin. Exp. Immunol.*, 26:181–187.
31. Goldring O. L., Sher, A., Smithers, S. R., and McLaren, D. J. (1971): Host antigens and parasite antigens of murine. *S. mansoni*. *Trans. R. Soc. Trop. Med. Hyg.*, 71:144–148.
32. Haas, W., and Schmitt, R. (1978): Chemical stimuli for penetration of *S. mansoni* cercariae. *Naturwissenschaften*, 65:110.
33. Hang, L. M., Warren, K. S., and Boros, D. L. (1974): *S. mansoni*: Antigenic secretions and the etiology of egg granulomas in mice. *Exp. Parasitol.*, 35:288–298.
34. Hansen, E. L. (1975): A cell line from embryos of *B. glabrata* (Pulmonata): Establishment and characteristics. In: *Invertebrate Tissue Culture. Research Applications*, edited by K. Maramorosch, pp. 75–99. Academic Press, New York.

35. Hockley, D. J., and McLaren, D. J. (1973): *S. mansoni*: Changes in the outer membrane of the tegument during development from cercaria to adult worm. *Int. J. Parasitol.*, 3:13–25.
36. Hockley, D. J., McLaren, D. J., Ward, B. J., and Nermut, M. V. (1975): A freeze fracture study of the tegumental membrane of *S. mansoni* (Platyhelminthes: Trematoda). *Tissue Cell*, 7:485–496.
37. Howells, R. E., Ramalho-Pinto, F. J., Gazzinelli, G., de Oliveira, C. C., Figueiredo, E. A., and Pellegrino, J. (1974): *S. mansoni*: mechanism of tail loss and its significance to host penetration. *Exp. Parasitol.*, 36:373–385.
38. Imperia, P. S., and Fried, B. (1980): Pheromonal attraction of *S. mansoni* females toward males in the absence of worm-tactile behavior. *J. Parasitol.*, 66:682–684.
39. Jordan, P., and Webbe, G. (1969): *Human Schistosomiasis*. Heinemann, London.
40. Kruger, S. P., Heitman, L. P., Van Wyk, J. A., and McCully, R. M. (1969): The route of migration of *S. matthei* from lungs to liver in sheep. *J. S. Afr. Vet. Med. Assoc.*, 40:39–43.
41. Kusel, J. R., and Mackenzie, P. E. (1975): The measurement of the relative turnover rates of proteins of the surface membranes and other fractions of *S. mansoni* in culture. *Parasitology*, 71:261–273.
42. Lawson, J. R., and Wilson, R. A. (1980): Metabolic changes associated with the migration of the schistosomulum of *S. mansoni* in the mammal host. *Parasitology*, 81:325–336.
43. Lawson, J. R., and Wilson, R. A. (1980): The survival of the cercariae of *S. mansoni* in relation to water temperature and glycogen utilization. *Parasitology*, 81:337–348.
44. Mahmoud, A. A. F., Warren, K. S., and Peters, P. A. (1975): A role for the eosinophil in acquired resistance to *S. mansoni* infection as determined by antieosinophil serum. *J. Exp. Med.*, 142:805–813.
45. McInnis, A. J. (1969): Identification of chemicals triggering cercarial penetration responses of *S. mansoni*. *Nature*, 224:1221–1222.
46. McLaren, D. J. (1980): *S. mansoni: The Parasite Surface in Relation to Host Immunity*. Research Studies Press, New York.
47. McLaren, D. J., Clegg, J. A., and Smithers, S. R. (1975): Acquisition of host antigens by young *S. mansoni* in mice: Correlation with failure to bind antibody in vitro. *Parasitology*, 70:67–75.
48. McLaren, D. J., and Hockely, D. J. (1977): Blood flukes have a double outer membrane. *Nature*, 279:147–149.
49. Miller, P., and Wilson, R. A. (1978): Migration of the schistosomula of *S. mansoni* from skin to lungs. *Parasitology*, 77:281–302.
50. Miller, P., and Wilson, R. A. (1980): Migration of the schistosomula of *S. mansoni* from the lungs to the hepatic portal system. *Parasitology*, 80:267–288.
51. Minard, P., Dean, D. A., Jacobson, R. H., Vannier, W. E., and Murrell, K. D. (1978): Immunization of mice with cobalt-60 irradiated *S. mansoni* cercariae. *Am. J. Trop. Med. Hyg.*, 27:76–86.
52. Moser, G., Wassom, D., and Sher, A. (1980): Studies of the antibody-dependent killing of schistosomula of *S. mansoni* employing haptenic target antigens: I. Evidence that the loss in susceptibility of immune damage undergone by developing schistosomula involves a change unrelated to the masking of parasite antigens by host molecules. *J. Exp. Med.*, 152:41–53.
53. Perez, H., and Terry, R. J. (1973): The killing of adult *S. mansoni* in vitro in the presence of antisera to host antigenic determinants and peritoneal cells. *Int. J. Parasitol.*, 3:499–503.
54. Phillips, S. M., and Colley, D. G. (1978): Immunologic aspects of host responses to schistosomiasis: Resistance, immunopathology, and eosinophil involvement. *Prog. Allergy*, 24:49–182.
55. Playfair, J. H. L. (1978): Effective and ineffective immune responses to parasites: Evidence from experimental models. *Curr. Top. Microbiol. Immunol.*, 80:37–64.
56. Samuelson, J. C., Sher, A., and Caulfield, J. P. (1980): Newly transformed schistosomula spontaneously lose surface antigens and C3 acceptor sites during culture. *J. Immunol.*, 124:2055–2057.
57. Schubert, M. (1948): Conditions for drug testing in experimental schistosomiasis mansoni in mice. *Am. J. Trop. Med.*, 28:121–136.
58. Shaw, J. R., Marshall, I., and Erasmus, D. A. (1977): *S. mansoni*: In vitro stimulation of vitelline cell development by extracts of male worms. *Exp. Parasitol.*, 42:14–20.
59. Sher, A., Hall, B. F., and Vadas, M. A. (1978): Acquisition of murine major histocompatibility complex gene products by schistosomula of *S. mansoni*. *J. Exp. Med.*, 148:46–57.
60. Shiff, C. J., Cmelik, S. H. W., Ley, H. E., and Kriel, R. L. (1972): The influence of human skin lipids on the cercarial penetration responses of *S. hematobium* and *S. mansoni*. *J. Parasitol.*, 58:476–480.

61. Smithers, S. R., and Terry, R. J. (1969): The immunology of schistosomiasis. *Adv. Parasitol.*, 7:41–93.
62. Smithers, S. R., and Terry, R. J. (1969): Immunity in schistosomiasis. *Ann. N.Y. Acad. Sci.*, 160:826–840.
63. Smithers, S. R., and Terry, R. J. (1976): The immunology of schistosomiasis. *Adv. Parasitol.*, 14:399–422.
64. Smithers, S. R., Terry, R. J., and Hockley, D. J. (1969): Host antigens in schistosomiasis. *Proc. R. Soc. Biol.*, 171:483–494.
65. Stirewalt, M. A. (1974): *S. mansoni*: Cercaria to schistosomule. *Adv. Parasitol.*, 12:115–182.
66. Stirewalt, M. A., and Austin, B. E. (1973): Collection of a secreted protease from the preacetabular glands of cercariae of *S. mansoni*. *J. Parasitol.*, 59:741–743.
67. Stirewalt, M. A., and Kruidenier, F. J. (1961): Activity of the acetabular secretory apparatus of cercariae of *S. mansoni* under experimental conditions. *Exp. Parasitol.*, 11:191–211.
68. Stirewalt, M. A., and Walters, M. (1973): Histochemical analysis of the postacetabular gland secretion of cercariae of *S. mansoni*. *Exp. Parasitol.*, 33:56–72.
69. Sturrock, R. F. (1966): Daily egg output of schistosomes. *Trans. R. Soc. Trop. Med. Hyg.*, 60:139–140.
70. Torpier, G., Capron, M., and Capron, A. (1977): Structural changes of the tegumental membrane complex in relation to developmental stages of *S. mansoni* (Platyhelminthes: Trematoda). *J. Ultrastruct. Res.*, 61:309–324.
71. Warren, K. S., Domingo, E. O., and Cowun, R. B. T. (1967): Granuloma formation around schistosome eggs as a manifestation of delayed hypersensitivity. *Am. J. Pathol.*, 51:735–756.
72. Warren, K. S., Mahmoud, A. A. F., Cummings, P., Murphy, D. J., and Mouser, D. B. (1974): Schistosomiasis mansoni in Yemeni in California: duration of infection, presence of disease and therapeutic management. *Am. J. Trop. Med. Hyg.*, 23:902–909.
73. WHO Memorandum (1974): Immunology of schistosomiasis. *Bull. WHO*, 51–:553–595.
74. Wilks, N. E. (1967): Lungs to liver migration of schistosomes in the laboratory mouse. *Am. J. Trop. Med. Hyg.*, 16:599–605.
75. Wilson, R. A. (1980): Is immunity to *S. mansoni* in the chronically infected laboratory mouse an artefact of pathology? *Proceedings of the Third European Multicolloquium Parasitology, Cambridge, England*, p. 37.

Molecular Biology of Parasites, edited by
J. Guardiola, L. Luzzatto, and W. Trager.
Raven Press, New York © 1983.

Molecular Biology of Antigenic Variation in African Trypanosomes

A. Bernards and P. Borst

Section for Medical Enzymology and Molecular Biology, Laboratory of Biochemistry, University of Amsterdam, Jan Swammerdam Institute, 1005 GA Amsterdam, The Netherlands

Trypanosomes are flagellate protozoa responsible for some of the more serious parasitic diseases of man and domestic animals (11). Among the mechanisms that parasites have evolved to evade immunodestruction, antigenic variation by African trypanosomes stands out as one of the more sophisticated (reviewed in refs. 9, 15, and 19). Antigenic variation involves the sequential synthesis of radically different variant surface glycoproteins (VSGs) that form a surface coat completely covering the trypanosome during its stay in the mammalian bloodstream (6,8). The size of the repertoire of VSGs that a trypanosome can express has been shown to be at least 101 in a clone of *Trypanosoma equiperdum* (7). To study the molecular biology of antigenic variation we—and several other groups—have cloned DNAs complementary to the messenger RNAs (mRNAs) of different VSGs of *Trypanosoma brucei* (1,12,16,20) or *T. equiperdum*. The trypanosome variants used in our work, designated variants 117, 118, 121, and 221 of *T. brucei* stock 427, were isolated from a chronically infected rabbit or during subsequent *in vitro* culture (12). Pure variants were obtained by inoculating single trypanosomes in irradiated mice.

VSG SYNTHESIS

Sequence analysis of complementary DNA (cDNA) clones has revealed that VSGs are synthesized with both an N- and a C-terminal hydrophobic peptide extension (3,4,14). An N-terminal signal peptide is to be expected on a protein made for export purposes; the C-terminal extension, on the other hand, may function as a "stop transfer" signal to guide the VSGs to their position on the outside of the cell membrane.

An unusual feature of VSG mRNAs is that they lack the polyadenylation signal found in other eukaryotic mRNAs, although VSG mRNAs do contain a poly(A) tail. Whether this absence contributes to the observed heterogeneity of the poly(A) addition site (14) remains to be determined. Another remarkable feature of VSG mRNAs is that the 3′ untranslated region of the mRNAs is rigorously conserved, even if the coding regions show no homology (14). This is explained by the special role played by this region in VSG gene activation, and we return to this point below.

VSG GENE ACTIVATION

Using the cloned cDNAs as hybridization probes in RNA blotting experiments, it was found that regulation of VSG synthesis occurs at the level of DNA transcription, since cDNA probes only detect an mRNA in the variant expressing the corresponding VSG (1,13). Moreover, DNA blotting experiments in which the cDNAs were used as probes of gene structure showed that the activation of some VSG genes is accompanied by the duplication of a preexisting invariable basic copy (BC) of the VSG gene, followed by the transposition of the duplicated gene to a putative expression site somewhere else in the trypanosome genome (5,13). This second copy of the VSG gene, the expression-linked copy (ELC), is only detected in DNA from the variant that actively expresses this gene, and it disappears when the trypanosome clone switches to the production of other VSGs (15a).

The construction of physical maps of the BC and ELC of VSG genes 117 and 118 (5,10,22) has revealed that the segment of the BC that is transposed to the expression site includes DNA sequences 1 to 2 kilobase pairs (kb) in front of the genes. Comparison of the physical maps of the 117 and 118 ELC shows that these two VSG genes are found in the same (or a very similar) "expression site," since the physical maps outside the transposed segment are identical for both genes (22). The general features of this mechanism have recently been confirmed by Pays et al. (17,18).

GENERATION OF THE ELC: A RECOMBINATION ALTERING THE 3′-END OF THE GENE

While the 5′ breakpoint of VSG gene duplication lies well in front of the gene, the 3′ breakpoint is situated in the part of the gene that codes for the C-terminal hydrophobic protein extension (2). This result has been obtained through sequence comparison of 117 cDNA and a cloned 117 BC, S_1 nuclease protection experiments and genomic blotting experiments. The latter approach showed that a BspI restriction endonuclease recognition site present in the 3′-terminal region of 118 cDNA but absent from the 118 BC gene is also present in the 118 ELC gene, confirming that the ELC is the active gene used for mRNA production. The sequence comparison of the 3′-terminal region of the 117 cDNA and the 117 cloned BC gene showed that these regions differ by multiple point mutations, small deletions, and insertions, the differences starting within the region that codes for the C-terminal hydrophobic peptide extension. Somewhat surprisingly, the 117 BC still seems to be a functional gene, since it includes an (altered) sequence coding for a hydrophobic peptide extension, a stop codon at the same position as in the cDNA, and all of the elements found in common between the 3′-untranslated regions of VSG mRNAs. This is not the result expected if a functional end of a VSG gene is created only during transposition by recombination with a 3′-end in the expression site, and it indicates that BC genes are somehow regularly checked for functionality. This could occur by a mechanism in which BC genes themselves are regularly activated, or alternatively, if the recombination that integrates the 3′-end of the duplicated BC gene in the expression site could occur anywhere in the last 150 nucleotides of the gene. This latter hypothesis is supported by a recent experiment in which it was shown that a newly isolated trypanosome clone that expresses the VSG 118 gene contains a 118 ELC with the same 3′-end as the 118 BC gene (15a).

NOT ALL VSG GENES ARE ACTIVATED BY GENE DUPLICATION-TRANSPOSITION

While the duplication-transposition mechanism holds for the VSG 117 and 118 genes and also for the AnTat 1.1 and 1.8 genes studied by Pays and co-workers (17,18), it is apparently not the only mechanism to activate VSG genes. Williams and co-workers (21) have described genomic rearrangements around the ILTat 1.2 and ILTat 1.3 VSG genes in expressor and nonexpressor trypanosome clones. Activation of these genes does not appear to involve a duplication-transposition. We have obtained analogous results with the VSG 221 gene in our stock. These aberrantly behaving VSG genes are always detected very early in chronic trypanosome infections. How these genes are activated remains to be determined.

ACKNOWLEDGMENTS

This work was supported in part by a grant to P.B. from the Foundation for Fundamental Biological Research (BION), which is subsidized by The Netherlands Organization for the Advancement of Pure Research (ZWO), and by financial support from the UNDP World Bank WHO Special Programme for Research and Training in Tropical Diseases [No. (TRY) T16/181/T7(34)].

REFERENCES

1. Agabian, N., Thomashow, L., Milhausen, M., and Stuart, K. (1980): Structural analysis of variant and invariant genes in trypanosomes. *Am. J. Trop. Med. Hyg., (Suppl.)* 29:1043–1049.
2. Bernards, A., Van der Ploeg, L. H. T., Frasch, A. C. C., and Borst, P. (1981): Activation of trypanosome surface glycoprotein genes involves a gene duplication-transposition leading to an altered 3′-end. *Cell*, 27:497–505.
3. Boothroyd, J. C., Cross, G. A. M., Hoeijmakers, J. H. J., and Borst, P. (1980): A variant surface glycoprotein of *Trypanosoma brucei* synthesized with a C-terminal hydrophobic tail absent from purified glycoprotein. *Nature*, 288:624–626.
4. Boothroyd, J. C., Paynter, C. A., Cross, G. A. M., Bernards, A., and Borst, P. (1981): Variant surface glycoproteins of *Trypanosoma brucei* are synthesized with cleavable hydrophobic sequences at the carboxy and amino termini. *Nucleic Acids Res.*, 9:4735–4743.
5. Borst, P., Frasch, A. C. C., Bernards, A., Van der Ploeg, L. H. T., Hoeijmakers, J. H. J., Arnberg, A. C., and Cross, G. A. M. (1981): DNA rearrangements involving the genes for variant antigens in *Trypanosoma brucei*. *Cold Spring Harbor Symp. Quant. Biol.*, 45:935–943.
6. Bridgen, P. J., Cross, G. A. M., and Bridgen, J. (1976): N-terminal amino acid sequences of variant-specific surface antigens from *Trypanosoma brucei*. *Nature*, 263:613–614.
7. Capbern, A., Giroud, C., Baltz, T., and Mattern, P. (1977): *Trypanosoma equiperdum:* Etudes de variations antigéniques au cours de la trypanosomiasis expérimentale du lapin. *Exp. Parasitol.*, 42:6–13.

8. Cross, G. A. M. (1975): Identification, purification and properties of clone-specific glycoprotein antigens constituting the surface coat of *Trypanosoma brucei. Parasitology*, 71:393–417.
9. Cross, G. A. M. (1978): Antigenic variation in trypanosomes. *Proc. R. Soc. London, Ser. B*, 202:55–72.
10. Frasch, A. C. C., Bernards, A., Van der Ploeg, L. H. T., Borst, P., Hoeijmakers, J. H. J.,Van den Burg, J., and Cross, G. A. M. (1980): The genes for the variable surface glycoproteins of *Trypanosoma brucei*. In: *The Biochemistry of Parasites and Host-Parasite Relationships: The Host-Invader Interplay*, edited by H. Van den Bossche, pp. 235–239. Elsevier/North-Holland, Amsterdam.
11. Hoare, C. A., editor (1972): *The Trypanosomes of Mammals: A Zoological Monograph*. Blackwell, Oxford.
12. Hoeijmakers, J. H. J., Borst, P., Van den Burg, J., Weissmann, C., and Cross, G. A. M. (1980): The isolation of plasmids containing DNA complementary to messenger RNA for variant surface glycoproteins of *Trypanosoma brucei. Gene*, 8:391–417.
13. Hoeijmakers, J. H. J., Frasch, A. C. C., Bernards, A., Borst, P., and Cross, G. A. M. (1980): Novel expression-linked copies of the genes for variant surface antigens in trypanosomes. *Nature*, 284:78–80.
14. Majumder, H. K., Boothroyd, J. C., and Weber, H. (1981): Homologous 3′-terminal regions of mRNAs for surface antigens of different antigenic variants of *Trypanosoma brucei. Nucleic Acids Res.*, 9:4745–4753.
15. Marcu, K. B., and Williams, R. O. (1981): Microbial surface elements: The case of variant surface glycoprotein (VSG) genes of African trypanosomes. In: *Genetic Engineering*, Vol. 3, edited by S. K. Setlow and A. Hollaender, pp. 129–155. Plenum, New York.
15a. Michels, P. A. M., Bernards, A., Van der Ploeg, L. H. T., and Borst, P. (1982): Characterization of the expression-linked gene copies of variant surface glycoprotein 118 in two independently isolated clones of *Trypanosoma brucei. Nucleic Acids Res.*, 10:2353–2365.
16. Pays, E., Delronche, M., Lheureux, M., Vervoort, T., Bloch, J., Cannon, F., and Steinert, M. (1980): Cloning and characterization of DNA sequences complementary to messenger ribonucleic acids coding for the synthesis of two surface antigens of *Trypanosoma brucei. Nucleic Acids Res.*, 8:5965–5981.
17. Pays, E., Van Meirvenne, N., LeRay, D., and Steinert, M. (1981): Gene duplication and transposition linked to antigenic variation in *Trypanosoma brucei. Proc. Natl. Acad. Sci. USA*, 78:2673–2677.
18. Pays, E., Lheureux, M., and Steinert M. (1981): The expression-linked copy of surface antigen gene in *Trypanosoma* is probably the one transcribed. *Nature*, 292:265–267.
19. Vickerman, K. (1978): Antigenic variation in trypanosomes. *Nature*, 273:613–617.
20. Williams, R. O., Young, J. R., and Majiwa, P. A. O. (1979): Genomic rearrangements correlated with antigenic variation in *Trypanosoma brucei. Nature*, 282:847–849.
21. Williams, R. O., Young, J. R., Majiwa, P. A. O., Doyle, J. J., and Shapiro, S. Z. (1981): Contextural genomic rearrangements of variable antigen genes in *Trypanosoma brucei. Cold Spring Harbor Symp. Quant. Biol.*, 45:945–949.
22. Van der Ploeg, L. H. T., Bernards, A. Rijsewijk, F. A. M., and Borst, P. (1982): Characterization of the DNA duplication-transposition that controls the expression of two genes for variant surface glycoproteins in *Trypanosoma brucei. Nucleic Acids Res.*, 10:593–607.

Molecular Biology of Parasites, edited by
J. Guardiola, L. Luzzatto, and W. Trager.
Raven Press, New York © 1983.

A Puzzle Genome: Kinetoplast DNA

Piero A. Battaglia, Marina del Bue, Marco Ottaviano, and Marta Ponzi

Istituto Superiore di Sanitá, 00161 Rome, Italy

Kinetoplast DNA (K-DNA) represents an original and peculiar structure of genome organization. We are familiar with two widespread and generalized genome organizations that have met with evolutionary success: (a) the genome structure organized in linear chromosomes, common to eukaryotic genomes; and (b) the circular structure peculiar to plasmids, viruses, bacteria, and mitochondria.

Intermediate structures exist between these two forms (or dynamic stages of genome organization). Examples are the "amplification" mechanism found in *Amphybian* and the "magnification" found in *Drosophila*. In both cases the rDNA genes are present in both linear and circular forms (25,36).

Today, another form of genome organization has come to light: the "fragmented genome" of macronuclei in ciliata. In these protozoa the macronuclear DNA is constituted by "free" genes organized in discrete classes of low molecular weight ranging from 0.5 to 20 kilobase pairs (kb) in length (18).

Among all these genome organizations, the K-DNA of Kinetoplastidae protozoa is the most "bizarre" (Fig. 1). K-DNA consists, in fact, of a large "tridimensional" network composed of thousands of catenated circles. It is localized in an organelle (kinetoplast) situated inside of the single mitochondrion of these protozoa.

In this paper we deal with the structure of the two K-DNA components: minicircles and maxicircles. We review the data on maxicircle function and analyze the minicircle on the basis of available results. Finally, we present a model K-DNA replication mechanism that refers to the minicircle component. Particular emphasis will be given to the model of structural organization of minicircles as revealed by sequence analysis and to its implications for the K-DNA function.

STRUCTURE

The kinetoplast network consists of a "major" component; the catenated circles (minicircles) comprising 95 to 98% of K-DNA, and a "minor" component, representing 2 to 5% of the total K-DNA. A single network contains approximately 5 to 20 $\times$ 10^3 minicircles and 20 to 50 maxicircles. The minicircle size varies from 0.9 to 2.5 kb in different species. The maxicircle size varies from about 20 to 30 kb (for reviews, see refs. 2,4,12,16).

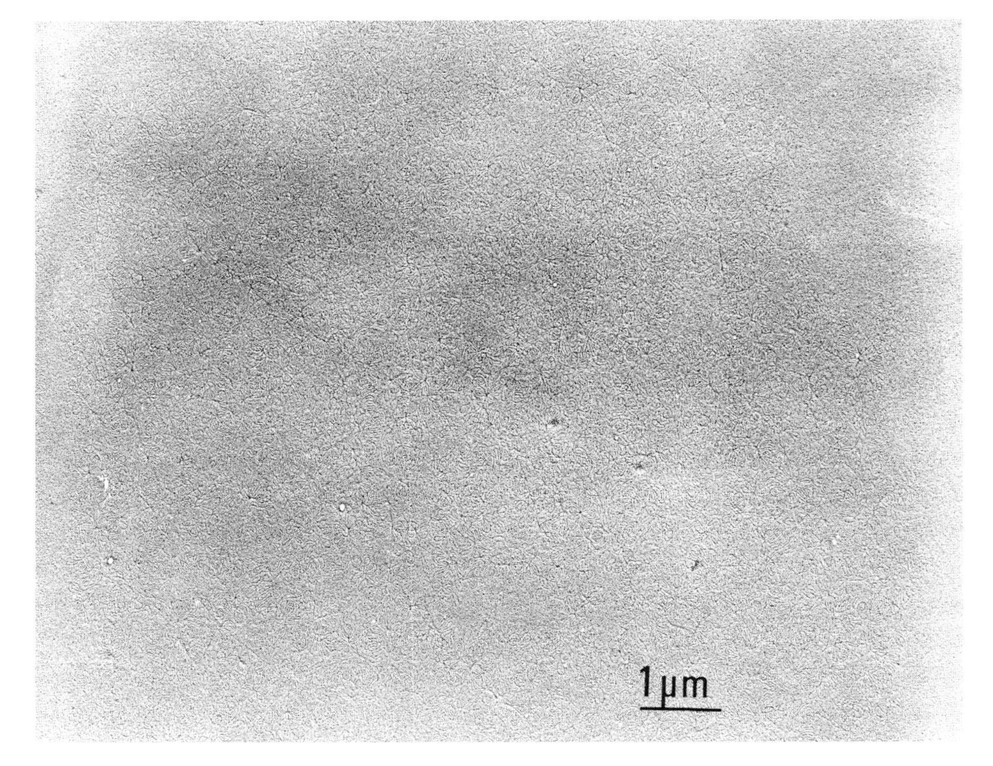

FIG. 1. Electron micrograph of purified K-DNA network from *T. lewisi.*

Minicircle Structure

The approximately 10,000 minicircles constituting most of the K-DNA network of Trypanosomatidae are exceptionally constant in length within a single species, although the length varies among species (Fig. 2A). This fact suggests that the enormous K-DNA network is constituted by circles homogeneous in base sequence, as would occur by replication of a single minicircle.

To test this hypothesis it is sufficient to analyze and compare the sequences of different minicircles. Single minicircles can be easily isolated and then amplified for further analysis with genetic engineering techniques. Chen and Donelson (8) cloned and sequenced two different minicircles of *Trypanosoma brucei*. Comparing the minicircle sequences, they observed that they were made up of a region of extensive sequence homology (constant region) and a region of very different sequences (variable region) (8). The same result was obtained by Simpson in *Leishmania tarantolae* (29).

It appears, then, that cloned minicircles are organized as shown in Fig. 2B, where a quarter of the length is occupied by the constant region and the remainder by the variable region. One can ask whether all the minicircles of a K-DNA of one species

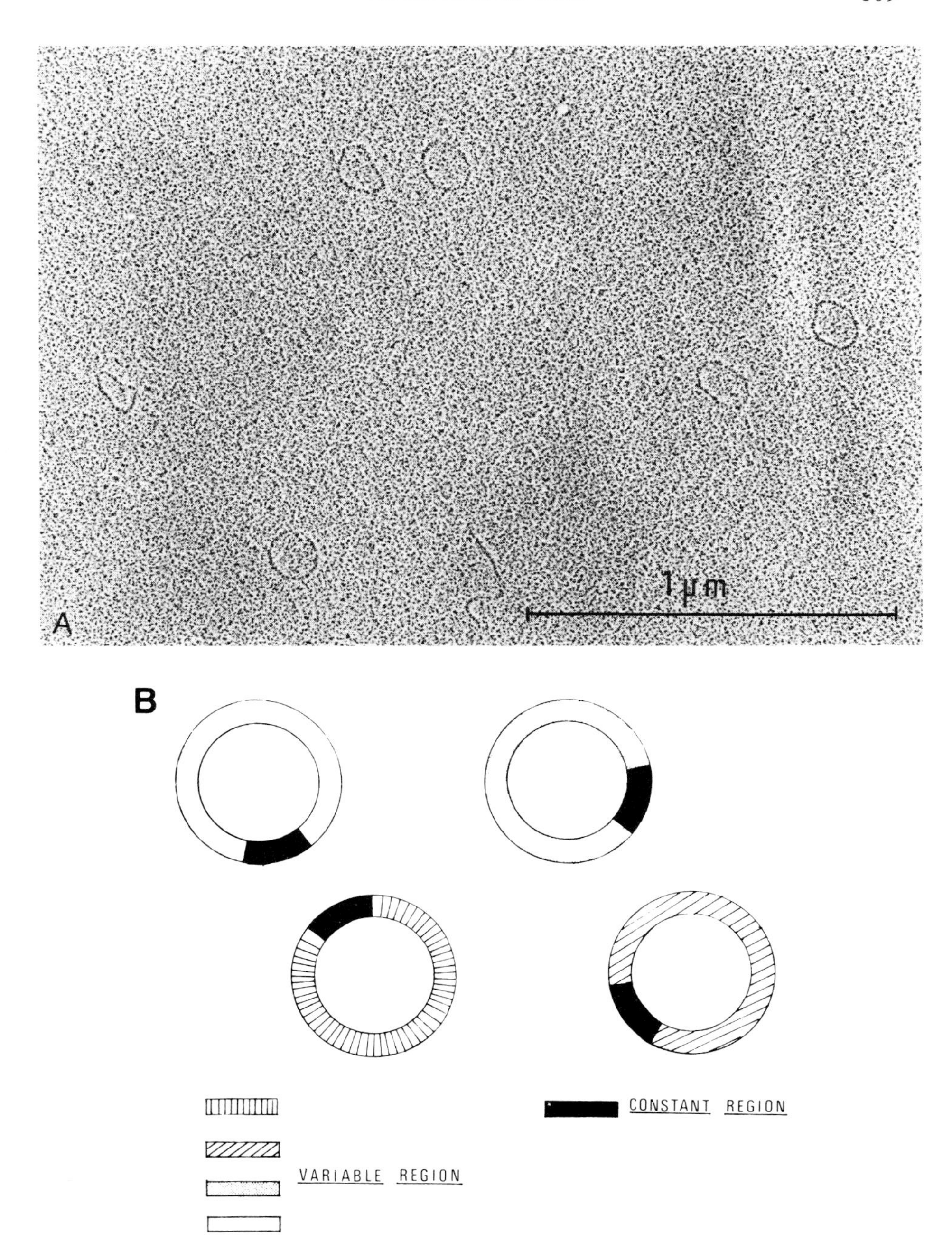

FIG. 2. A: Electron micrograph of free minicircles from K-DNA of *T. lewisi*. The contour length (0.4 m) of minicircles is homogeneous. **B:** Model of minicircle structure. See the text for details.

have the same organization and whether this organization is extended to other species.

Constant Region

A constant region seems to be present in all the minicircles of *T. brucei* (9), as is true in *L. tarantolae* (32). As a matter of fact, in these two cases all minicircles cloned as recombinant DNA in *Escherichia coli* hybridize to some extent within each species.

We observed that in *T. lewisi* (Fig. 3) different minicircle clones hybridize with each other, thus demonstrating the presence of a constant region. This is constituted by sequences of different length in different species. In *T. brucei* it is 120 base pairs (bp) long (minicircles are 1,000 bp), 91 bp in *L. tarantolae* (in which minicircles are 720 bp), and 1,400 bp in *Crithidia* (in which minicircles are 2,400 bp). The length of the constant region, therefore, does not seem to be proportional to the minicircle length. It seems to be highly conserved in nucleotide sequence within the same K-DNA but very different across species. Simpson (9) observed a dif-

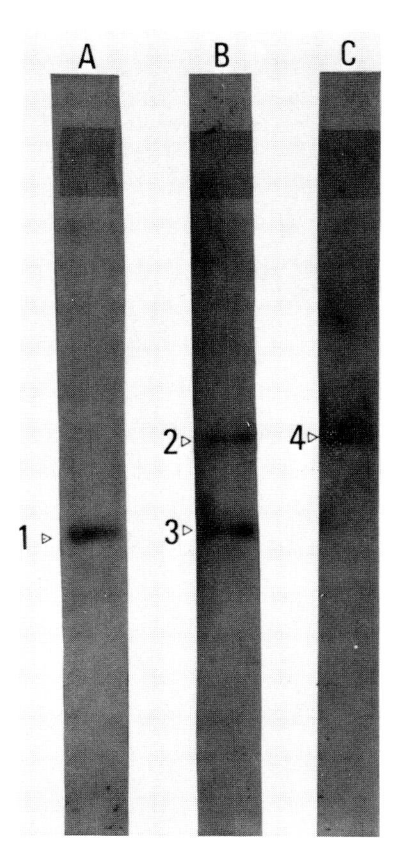

FIG. 3. Hybridization of nick-translated minicircles cut by Hind III probe with total digested K-DNA or cleaved cloned minicircle sequences. **A:** Hybridized linearized minicircle cloned in Bam H_1 site of pBR 322 (1). **B:** Total K-DNA cleaved by Hind III: hybridized band corresponding to minicircles dimers (2) and to linearized minicircles (3). **C:** Hybridization band of Hind III minicircle dimer cloned in pBR 322 (4).

ference of only 4 nucleotides out of the 91 constituting the constant region in *L. tarantolae*. In *T. brucei* the difference is of 8 nucleotides among the 122 constituting the constant region (8). Comparison of the constant regions of these two species shows only small homologous blocks of nucleotides (not more than 10 to 15 nucleotides in length), whereas the remaining sequences are different.

Of much interest in terms of possible functional implications is the fact that in the constant region a base sequence is present that can be transcribed in a polypeptide beginning with a starting codon (open reading frame). In *L. tarantolae* (32) the open reading frame is entirely contained in the constant region; in *T. brucei* (8) the starting codon and a long base sequence are also in the constant region, while the reading continues in the variable region.

Since the existence of a constant region was demonstrated in salivary *(T. brucei)* and stercoral *(T. lewisi)* trypanosomes, phylogenetically distant, and in *L. tarantolae* and *Crithidia*, different genuses of Trypanosomatidae, it seems to be a conserved evolutionary structure present in the whole biological order.

Variable Region

An exclusive merit of the minicircle model (Fig. 2B) deduced by sequence analysis is that it rationalizes, in a unique scheme, the enormous amount of dissimilar data on the minicircle heterogeneity demonstrated in almost all Trypanosomatidae (for reviews, see refs. 2,4,12,16).

The extent of minicircle sequence heterogeneity can be estimated by the pattern of restricted K-DNA in agarose gel electrophoresis. The characteristic pattern generally observed in K-DNA is as follows: (a) only a minor proportion of the catenated minicircles is digested by a given restriction enzyme, whereas the major part remains at the top of the gel; this is because it does not contain sites for that particular enzyme. Digested minicircles give rise to a heterogeneous pattern comprising the unit length of linearized circles and fragments of circles (Fig. 4).

As indicated by the model, minicircle heterogeneity would be the result of the variable region of minicircles. The results obtained by Simpson (32) have shown that variable regions do not contain the same restriction sites. Chen and Donelson (8) found small homology regions in 880 nucleotides, without an obvious pattern of localization along the sequence. The longest one was 15 nucleotides long. Together these small regions represent 150 homologous base pairs out of 880. While the constant regions are homologous in more than 90% of their length, the variable ones differ from each other in more than 80% of theirs.

The number of different variable regions present in K-DNA minicircles of a given species can be estimated by measurements of genetic complexity (C_0t) (see C. Frontali, *this volume*). This analysis was performed by Donelson et al. (9) in *T. brucei*, and 300 different classes of variable regions were found. The number of classes seems to be larger in *T. brucei* (33), is intermediate in *Crithidia* (10 to 20 classes) (33) and in *T. mega* (70 classes) (33), and very low in *L. tarantolae* (3 classes) (7,14) and in *T. equiperdum* (26).

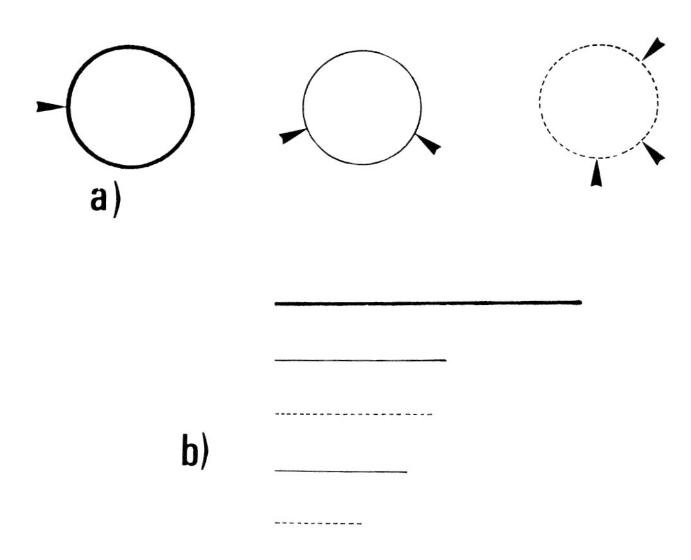

FIG. 4. **A:** Schematic representation of minicircle sequence heterogeneity (indicated by heavy, thin, and dotted rings) after restriction enzymes digestion. The arrows indicate the recognition site on different minicircles for the same restriction enzyme. **B:** Hypothetical restriction pattern after gel electrophoresis of the minicircle class shown in A. The heavy line indicates linearized minicircles, the thin and dotted lines indicate minicircles cut more times by the same restriction enzyme. The sum of the fragments overcomes the single unit length, showing the heterogeneity in base sequences.

The different classes of variable regions are present in the K-DNA network in different proportions. Challberg and Englund (7) therefore concluded that in *L. tarantolae* there are three major minicircle sequence classes in the ratio 7:1.5:0.7.

The number and relative proportion of different classes change over the time. Simpson (9) observed only two changed bands out of approximately 19 restriction fragments when he compared the restriction pattern of *L. tarantolae* obtained from a frozen strain and from the same strain kept in serial culture for four years. This rate of change agrees well with the results obtained in *Crithidia* (5) and in *T. cruzi* (15).

There are several possible mechanisms for the introduction of sequence diversity into K-DNA minicircle: (a) mutational or/and recombinant events, or (b) changes in the relative proportion of minicircle classes.

Maxicircle Structure

The earliest evidence that another component in addition to minicircles was present in purified K-DNA was obtained by Steinert and Van Assel in 1976 (34,35). They detected occasional 11 μm circles in electron micrographs of K-DNA from *Crithidia*. Kleisen and co-workers (19,20) showed that these circles were linked in a network by catenation. These structures have been called maxicircles.

Evidence of maxicircles in almost all Trypanosomatidae includes the appearance of high-molecular-weight fragments on agarose gel electrophoresis after restriction with several enzymes and the observation of supertwisted "edge loops" extending out from the periphery of the network made by electron microscopy (21). Figure 5 shows the long loops extending from a network K-DNA preparation of *T. lewisi*. The proportion of maxicircles present in a network is very low, about 3 to 20% of total K-DNA in Trypanosomatidae; their molecular weight ranges from 12×10^6 in *T. brucei* to 22×10^6 in *Crithidia*, as estimated in measurements by electron microscopy or gel electrophoresis.

Experimental data indicate that maxicircles are homogeneous in base sequence, in contrast to minicircles. Thus, analyzing the high-molecular-weight bands that appear on agarose gel after digestion with restriction endonucleases, it is observed that the sum of the fragments adds up to the molecular weight expected by the single unit.

REPLICATION OF K-DNA

Much effort has been spent in the last few years to elucidate the replication mechanism of the complex structure of K-DNA. The catenated structure of minicircles is maintained in replicating networks during the S phase of the cellular cycle,

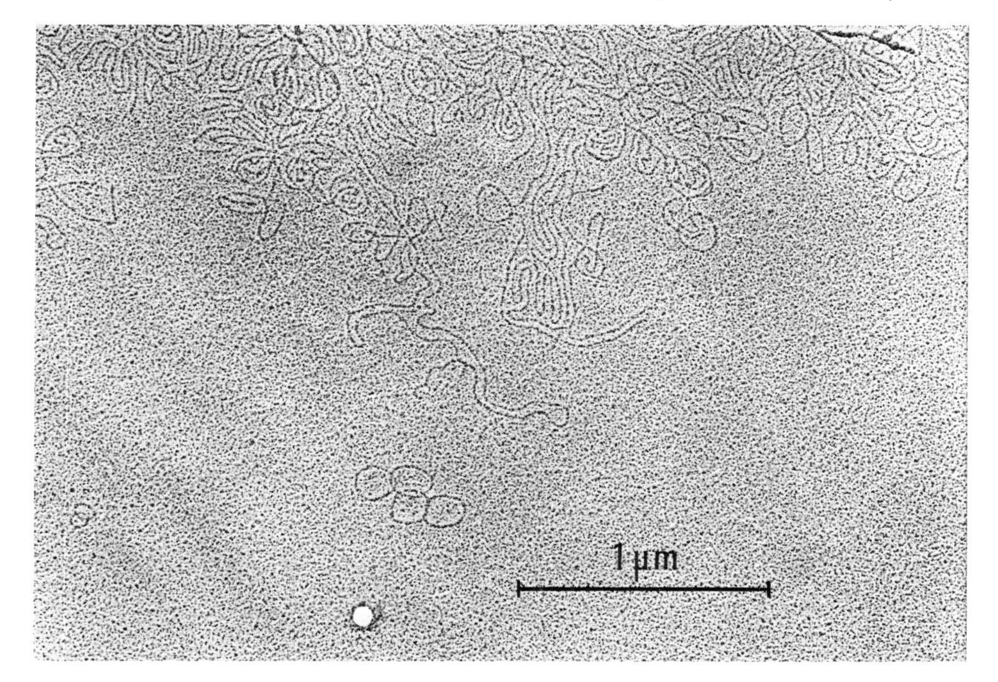

FIG. 5. Electron micrograph of free minicircles and maxicircles (long DNA loops extending at the edge of the network) in a K-DNA preparation of *T. lewisi*.

and a precise segregation of the two daughter networks occurs during cell division. Moreover, because of the presence of a single mitochondrion in Kinetoplastidae, the replication of mitochondrial and nuclear DNA has to be synchronous to guarantee the preservation of this organelle in each daughter cell.

Each minicircle must be considered as an independent replicon because of the circular structure and because it replicates only once for generation (22). Englund elaborated a replication model on the basis of experimental observations. He observed that in *Crithidia* a portion (about 0.4%) of minicircles was free from the network structure in exponentially growing cells (11). When [^{3}H]thymidine was added to the culture, 10% of the labeled precursor was found in this minicircle fraction after a 5-min pulse. He hypothesized that these labeled minicircles could be intermediates of replication. A hypothetical enzyme, e.g., topoisomerase, could randomly release minicircles from the network. They would replicate in this free form, and then another enzyme would reattach the progeny represented by nicked minicircles to the periphery of the replicating network. The migration of replicated minicircles toward the center of the network allows the release of another portion of the minicircles.

This idea was supported by autoradiographs of networks after pulse-labeling experiments with [^{3}H]thymidine precursor (29,30). The silver grains were localized in a ring around the periphery of the network. After a short chase, the ring of silver grains moved towards the center of the network. At the end of the replication process, the doubled network is constituted by nicked minicircles. The repair of the nicks and the segregation of the two progeny networks terminate the process.

Another support to the model comes from the characterization of K-DNA in exponentially growing cells by CsCl-propidium diiodide gradients (10). In this technique DNA is shown as a fluorescent band, and separating DNA components with different conformations (for example, nicked, supercoiled, and linear DNA) show up as different bands.

Using equilibrium centrifugation in a CsCl-propidium diiodide gradient and electron microscopic analysis, Englund (10) observed that the K-DNA of *Crithidia* could be separated into three different forms of network: form I containing covalently closed minicircles and representing a nonreplicated network; form II containing nicked and gapped minicircles and having about twice the molecular weight of form I; and form III, intermediate between the other two forms and containing some covalently closed minicircles and other that are nicked or gapped.

The role of these three networks in K-DNA replication process was studied in pulse-chase experiments. After a pulse of 10 min with [^{3}H]thymidine, radioactivity was found in a zone between forms I and II; after 30 min of chase, the radioactivity appeared in form II; after 60 min, in the form I; and after 180 minutes (nearly one generation time), some radioactivity reappeared in the intermediate fraction (10).

These experiments demonstrate that each form represents a particular stage of K-DNA at different times during the replication process. Form I is composed of unreplicated networks; form II is made of doubled networks, and it may represent the ultimate replication step in which all minicircles are replicated but still nicked; and form III is probably an intermediate form.

This model postulates the existence of two different enzymes that recognize replicated minicircles from those that are not. Englund suggested that the presence of nicks in replicated minicircles could be the feature by which the enzymes recognize the two different forms, allowing minicircles to be replicated only once per generation (13).

MINICIRCLE FUNCTION

In molecular biology it is very useful to isolate single biological units (genes and proteins) in order to study their function.

Genetic engineering techniques circle can easily isolate and then amplify a single minicircle for further analysis. Analyzing a cloned minicircle in heterologous systems, it is possible to study the replication mechanism, RNA transcription, and sequence expression in proteins.

Replication Region of Minicircles

Kidane and Simpson (personal communication) inserted single minicircles (pLT 19 and pLT 154) of *L. tarantolae* in the YIp5 plasmid of yeast, which has no replicon and therefore is unable to replicate in its natural host. In fact, alone, the YIp5 plasmid does not show autonomous replication and exhibits a low frequency of transformation. When a minicircle is inserted in YIp5, several transformants are obtained in yeast. This result demonstrates that a replicon is present in the inserted minicircle and that it is able to function in yeast.

To localize the minicircle region in which the replicon is contained, Kidane subcloned different DNA fragments of minicircles, obtained by restriction with different enzymes, in the YIp5 plasmid. He observed that the plasmid did not give transformants when the site for the restriction enzymes was located in the minicircle constant region, and therefore the recombinant plasmid contained only a part of this region. This result is consistent with the hypothesis that the replicon is localized in the constant region.

Possible Sequence Expression

K-DNA is the only known DNA in the trypanosome mitochondrion. As discussed below, good evidence exists that maxicircles code for the gene products that contribute to the biogenesis of mitochondria in all other organisms. Here, we propose some hypotheses concerning minicircle function; others are discussed in our concluding remarks.

Protein products codified by minicircles have yet to be demonstrated, but one can ask the following:

1. Do minicircles possess sequences that can be recognized by RNA polymerase as transcription initiation sites?

2. Are there minicircle sequences that can be correctly read to code for a sufficiently long protein?

Concerning the first question, our sequence analysis of the *T. lewisi* minicircle shows the presence of sequences in the same DNA fragment that are ordered in the same way as the prokaryotic promoters so far studied. The nucleotide sequence of the minicircle studied by us is also very similar to that of prokaryotic promoters (Fig. 6A,B).

Concerning the second question, we note that the minicircle sequence from *T. lewisi* contains a long open reading frame. Open reading frames were also observed by Chen, Donelson, and co-workers (8) in *T. brucei* and by Simpson and co-workers (32) in *L. tarantolae*. These sequences in the two minicircles of *T. brucei* are 156 and 213 nucleotides long. These data are therefore consistent with possible minicircle transcription and translation. How likely is this possibility?

Sequence Expression

With regard to the expression of minicircle DNA, preliminary evidence for the existence of a small RNA transcript in *Crithidia acantocephali* was found by Fouts (15), although such a transcript was not observed in Southern blotting experiments in *C. luciliae* (17).

To test whether minicircles can code for proteins *in vivo*, we transformed an *E. coli* strain producing minicells with the pBR 322 plasmid containing a minicircle cut by Bam HI restriction enzyme.

Minicells are small, spherical, anucleate bodies continuously produced during the growth of a mutant strain of *E. coli* K12, originally isolated and described by Adler et al. (1). Minicells purified from F-parental cells by sucrose-gradient sedimentation are deficient in chromosomal DNA but contain all the apparatus for protein biosynthesis. The demonstration that minicells containing plasmids are capable of incorporating ribonucleic acid and protein precursors into acid-precipitable material (28) indicated that they would be of use in examining *in vivo* RNA and protein synthesis in the absence of chromosomal macromolecular synthesis. *E. coli* minicells have been used to examine the RNA and protein synthesis directed by a variety of plasmids.

In our experiments, once-transformed minicells have been isolated and incubated with [^{35}S] methionine. Acrylamide gel electrophoresis and autoradiography reveal a band corresponding to a molecular weight of 19,000 (Fig. 7), a band that is absent in control samples.

FIG. 6. A: The three known promoter sequences are compared with a putative promoter found in the sequence of cloned minicircle of *T. lewisi.* The space between R_σ and R_c and between R_c and I usually contains 12 to 14 base pairs and 5 to 6 base pairs, respectively. **B:** Initiation at a promoter sequence. Initiation occurs in three general steps, mediated by contact at R_σ and R_c with initiation at I. **A:** Promoter recognition (recognition complex). **B:** Melting in (RS complex). **C:** Initiation at I, followed by chain elongation and sigma release. A nascent RNA molecule is included.

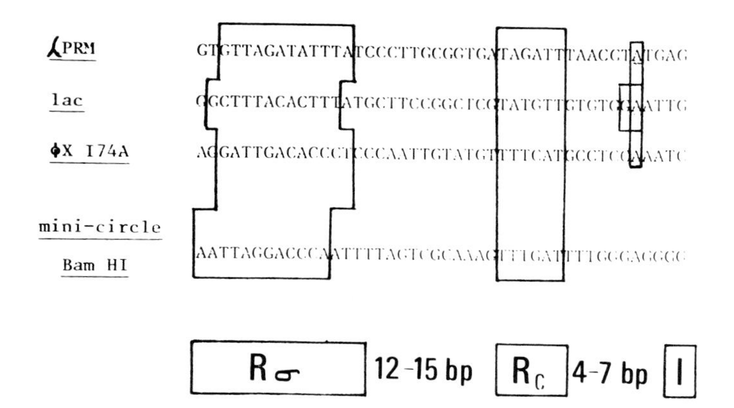

Rσ - σ recognition

Rc - core polymerase recognition

A **I** - initiation point(s)

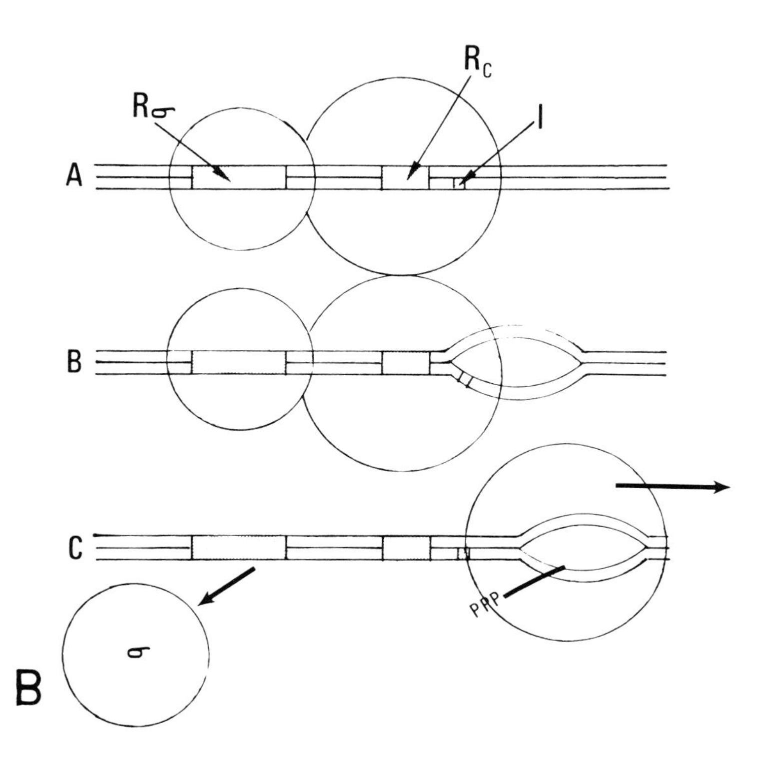

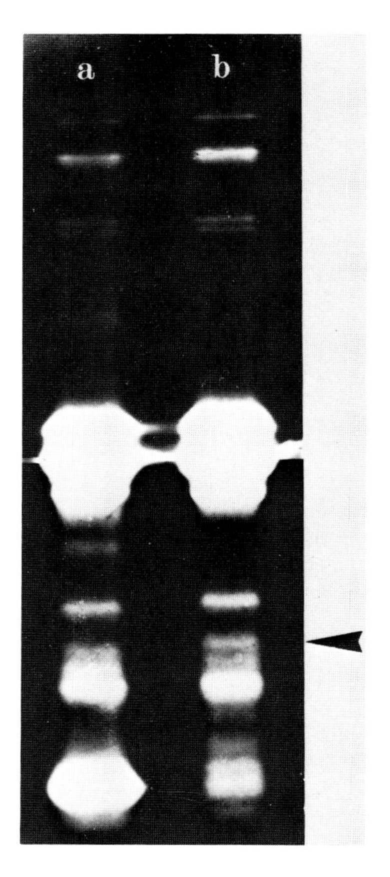

FIG. 7. Gel electrophoresis of the polypeptides synthesized in minicells containing pBR 322 plasmid **(a)** and minicircle cloned in Bam H1 site of pBR 322 **(b)**. The arrow indicates the position of a polypeptide of molecular weight of about 19,000, which is present in cloned minicircle and absent in pBR 322.

This result appears to confirm that *in vivo* the open reading frame present in *T. lewisi* effectively allows translation of the genetic information contained in the cloned minicircle into a polypeptide.

Two Extreme Hypotheses

Two unconventional hypotheses have been presented concerning the function of minicircles. They represent extreme hypotheses, and some of experiments performed to test them are reviewed here.

Minicircles as transposons

The first hypothesis was presented by Borst in a recent review (4), namely that minicircles are used as insertion sequences to switch genes on and off. This "transposon" function could be exercised on chromosomal and not on kinetoplast DNA. In effect, minicircles do not hybridize with the maxicircles. This hypothesis predicts the presence of inserted copies of minicircle DNA in the nucleus.

We tested this hypothesis as follows. A minicircle cut by Bam HI was cloned in the plasmid pBR 322. The cloned DNA was labeled at high specific radioactivity by nick translation with a radioactive nucleotide and hybridized to a cytological preparation of *T. lewisi* (Fig. 8).

The autoradiography of a cytological preparation of stained (Giemsa) trypanosomes shows the disintegration grains localized in the kinetoplast region of *T. lewisi*. Since, after a very long exposure, disintegration grains localized in the nucleus do not appear, we can estimate that sequences complementary to those cloned by us are not represented in the nucleus. Cytological hybridization therefore does not support the hypothesis proposed by Borst.

Mutants of K-DNA

Mutants of trypanosomes are known to have lost the ability to make functional mitochondria. Such mutants are found in nature, and they can be induced *in vitro* by intercalating dyes. The study of mutant trypanosomes with defective K-DNA may provide new information about the role of this K-DNA.

Such mutants found in nature are of the *T. brucei* group *(T. equiperdium, T. equinum, T. evansi)*. In these trypanosomes, different types of defects may exist:

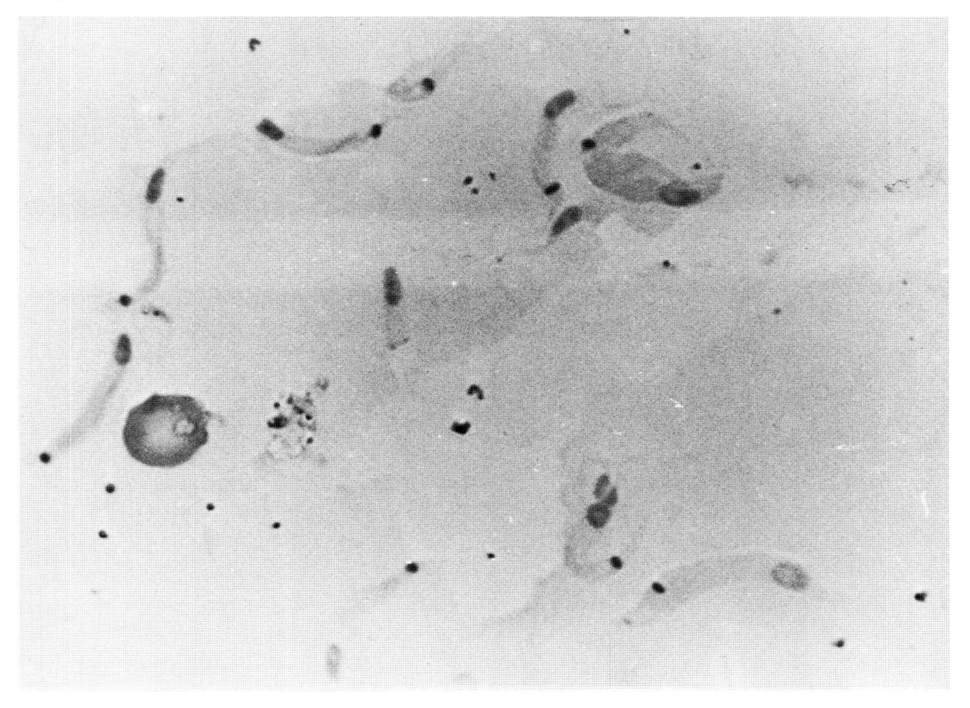

FIG. 8. *In situ* hybridization of [^{3}H]-ATP-labeled cloned minicircle with cytological preparation of trypanosomes. The labeled DNA was 7.4 × 10^6 cpm/mg.

(a) maxicircles are absent, but minicircles are present; (b) minicircles are absent, but maxicircles are present. In both cases, the trypanosomes have lost the ability to develop normal mitochondria (4).

Riou and co-workers (27) induced the loss of K-DNA in *T. equiperdium* in the bloodstream of infected rats by treatment with two intercalating drugs: ethidium bromide and acriflavine. K-DNA loss was complete, as shown by analytical ultracentrifugation in Cs/Cl/dye gradients and by reassociation kinetics of K-DNA labeled *in vitro*.

The dyskinetoplastic trypanosomes are viable in the bloodstream stages, and no difference in infectivity with respect to kinetoplastic trypanosomes was found.

From these data Riou concluded that no component of K-DNA network is essential to the viability of *T. equiperdium* in the bloodstream stage.

These data were interpreted as indicating that the K-DNA of trypanosomes is an example of "junk" DNA—functionless DNA that is conserved in the course of evolution in several prokaryotic and eukaryotic genomes (24). This interpretation is, nevertheless, inexact, since K-DNA could be dispensable in the bloodstream forms but not dispensable in invertebrate forms. This suggestion comes from the observation that natural mutants defective in K-DNA are not diffused by a vector insect but directly from one vertebrate to another. Before concluding that K-DNA is without function, we need to know whether dyskinetoplastic mutants can infect the insect vector and then the vertebrate.

FUNCTION OF MAXICIRCLES

The genetic function of K-DNA remains one of the unsolved problems about this bizarre DNA. No other DNA is present in the single mitochondrion, but which one of its two components (maxi- or minicircles) corresponds to mitochondrial DNA? When the maxicircles were discovered and it was found that they had highly conserved sequences and were comparable in size to mitochondrial DNA molecules (4), it seemed reasonable to suppose that they corresponded to the mtDNA.

The first approach to verify this hypothesis was to study the transcription of K-DNA. The 9 and 125 RNA of *L. tarantolae* (31), representing the major stable RNA species localized specifically in the kinetoplast fraction, hybridized selectively to the maxicircle sequence of the kinetoplast. Hoeijmakers and Borst (17) hybridized the total cellular RNA of *C. lucilia* with the K-DNA of this organism. They did not find minicircle transcripts under conditions wherein maxicircle fragments showed extensive and specific hybridization. They observed predominant hybridization with a segment of only 2,300 to 2,500 bp, from which it was inferred that this segment coded for unusually small mitochondrial ribosomal RNAs. In effect, it is known that all mtDNAs contain information for a specific ribosomal RNA.

The presence of ribosomal sequences on maxicircles makes it likely that maxicircles contain information typical of mitochondrial DNAs.

To pursue the study of the maxicircles transcripts, genetic engineering techniques were used. Borst (5) mapped RNA transcripts on *T. brucei* maxicircles using two restriction (Eco RI) fragments of maxicircles cloned in *E. coli*. This work showed that the two most prominent RNA species detected in DNA and RNA blots are the 9- and 12-S RNAs; six minor transcripts were identified on RNA blots. These minor RNAs were retained on digo (dt) cellulose and were presumably mitochondrial mRNAs or mRNA precursors containing a poly(A) tail, as also observed in the minor RNAs of animal mitochondria. He concluded that the transcripts found were of sufficient diversity to code for the unusual set of mitochondrial gene products.

The 9- and 12-S RNAs are ribosomal RNAs, as demonstrated by the fact that they were in about 1:1 stoichiometry, were not retained on oligo (dt) cellulose, had a size and sequence that was highly conserved among the Kinetoplastidae flagellates studied, and showed low but significant sequence homology with RNA genes of human mtDNA and ribosomal genes of *E. coli*. Similar results were obtained by Simpson (32) in *L. tarantolae*.

All these data establish that the 9- and 12-S RNAs are in effect rRNAs and that maxicircles code for at least one junction found on mtDNA in other organisms.

In the last few years, mitochondrial genomes have been studied intensively and are now well characterized. Some of them have been found in all mitochondria studied thus far (5). Proof that K-DNA contains structural mitochondrial genes was provided by Simpson et al. (personal communication), who found hybridization between structural yeast genes and *Leishmania* maxicircles.

In conclusion, the highly conserved sequences of maxicircles, the presence of ribosomal genes, and the cross-hybridization of maxicircles are consistent with the interpretation that they are the equivalent of the mitochondrial DNAs in other species.

CONCLUSIONS

The mitochondrial function of maxicircles seems to be well supported by experimental data: maxicircles have highly conserved sequences, contain ribosomal genes, and show homology with yeast structural genes.

Then what is the function of minicircles? Two general observations, supported by experimental data, can be made:

1. Each minicircle can be considered a single replication unit, since it replicates only once per generation and has a circular structure.
2. Replication is the only well-confirmed functional property of minicircles.

The first problem that arises is to establish the part of the minicircle structure in which the replication function is localized. The model of minicircle structure (presence of a constant and a variable region) supported by hybridization, cloning, and sequencing data suggests that the replication zone is localized within the constant region.

It is difficult to believe that the same function can be played by a region of variable base sequence within the same K-DNA—in other words, that different proteins (coded for by different sequences) can have the same function.

The strongest evidence supporting the fundamental role of the constant region in the replication process comes from the cloning experiments of different minicircle regions in plasmids unable to replicate (G. Kidane et al., *personal communication*). These preliminary experiments show that the constant region of the minicircles restores the missing sequences necessary for autonomous replication.

For several years minicircles have been considered as "silent" DNA; only recent evidence indicates that minicircles sequences can be expressed:

1. Fouts (15) has found that a portion of *Crithidia* minicircles hybridized to an RNA fraction.
2. Sequence analysis revealed an open reading frame of different length in three trypanosomes species.
3. We demonstrated that a cloned minicircle coded for a polypeptide.

Further experiments are required to answer the following questions:

1. Does the RNA hybridize to the constant or to the variable region of minicircles?
2. Does the synthesized protein correspond to that codified by the open reading frame?
3. What is the importance of these molecules in the replication process?

The following observations suggest to us an hypothesis on the role played by minicircles in K-DNA replication:

1. The number of minicircles in K-DNA of *T. brucei* varies at different stages of the cell cycle (23).
2. Minicircles are heterogeneous in base sequence, within the same species, but appear to be highly conserved in length. The selection therefore seems to operate so as to maintain a constant length of replication unit.

We propose that minicircles are elements regulating the replication rate of mitochondria during the cell cycle. The observed variation in number might be required to maintain an equal number of mitochondrial and nuclear replicons. Modification of minicircle length would imply a variation in replicon length, which would prevent the synchronization of nuclear and mitochondrial replication.

Because of the presence of a single mitochondrion per cell, the synchronization of the replication process guarantees the preservation of this organelle in each daughter cell. This unusual DNA function was described by Cavalier-Smith (6) in order to solve the DNA *c*-value paradox in eukaryotes.

Other hypotheses concerning the function of minicircles have been advanced:

1. Minicircle transcripts could be used in the processing of precursor RNAs, e.g., by forming duplexes with control segments of precursor RNAs (4).
2. Minicircle networks have a structural role in cell division, e.g., in the ordered division and segregation of the flagellum (35).

Further experimental work is required to establish the validity of these and other hypotheses.

In conclusion, maxicircles seem to be similar to mitochondrial DNA of the host; minicircles, on the other hand, possess a peculiar structure and function that differentiate them from the host mitochondrial DNA. This peculiar DNA could be a unique target, allowing a specific challenge to the parasite.

ACKNOWLEDGMENTS

This investigation received financial support from the UNDP/World Bank/WHO Special Programme for Research and Training in Tropical Diseases.

We would like to thank L. Verni for his expert technical assistance and T. Forte for her help in electron microscopy.

REFERENCES

1. Adler, H. I., Fisher, W. D., Cohen, A., and Hardigree, A. A. (1967): Miniature *Escherichia coli* cells deficient in DNA. *Proc. Natl. Acad. Sci. USA*, 57:321–326.
2. Barker, P. C. (1980): The ultrastructure of kinetoplast DNA with particular reference to the interpretation of dark field electron microscopy images of isolated, purified network. *Micron*, 11:21–62.
3. Borst, P., and Grivell, L. A. (1981): Small is beautiful—portrait of a mitochondrial genome. *Nature*, 162:27–28.
4. Borst, P., and Hoeijmakers, J. H. J. (1979): Kinetoplast-DNA. *Plasmid*, 2:20–40.
5. Borst, P., Hoeijmakers, J. H. J., Frasch, A. C. C., Sinijders, A., Janssen, J. W. G., and Fase-Fowler, F. (1980): The kinetoplast DNA of *Trypanosoma brucei*: Structure, evolution, transcription, mutants. In: *The Organization and Expression of the Mitochonrial Genome*, edited by A. N. Kroon and C. Saccone, pp. 7–19. Elsevier/North-Holland, Amsterdam.
6. Cavalier-Smith, T. (1978): Nuclear volume control by nucleoskeletal DNA, selection for cell volume and cell growth rate, and the solution of the DNA C-value paradox. *J. Cell. Sci.*, 34:247–278.
7. Challberg, S., and Englund, P. (1980): Heterogeneity of minicircles in kinetoplast DNA of *Leishmania tarentolae*. *J. Mol. Biol.*, 138:447–472.
8. Chen, K. K., and Donelson, J. E. (1980): Sequences of two kinetoplast DNA minicircles of *Trypanosoma brucei*. *Proc. Natl. Acad. Sci. USA*, 77:2445–2449.
9. Donelson, J. E., Majiwa, P. A. O., and Williams, R. O. (1979): Kinetoplast DNA minicircles of *Trypanosoma brucei* share regions of sequence homology. *Plasmid*, 2:572–588.
10. Englund, P. (1978): The replication of kinetoplast DNA networks in *Crithidia fasciculata*. *Cell*, 14:157–168.
11. Englund, P. (1979): Free minicircles of kinetoplast DNA in *Crithidia fasciculata*. *J. Biol. Chem.*, 254:4895–4900.
12. Englund, P. (1980): Kinetoplast DNA. In: *Biochemistry and Physiology of Protozoa*, 2nd ed., Vol. 4, edited by M. Landowski and S. H. Hutner, Academic Press, New York.
13. Englund, P., and Marini, J. C. (1980): The replication of kinetoplast DNA. *Am. J. Trop. Med. Hyg. (Suppl. 5)*, 29:1064–1069.
14. Englund, P., Di Maio, D., and Price, S. (1977): A nicked form of kinetoplast DNA in *Leishmania tarentolae*. *J. Biol. Chem.*, 252:6208–6216.
15. Fouts, D., and Wolstenholme, D. (1979): Evidence for a partial transcript of the small circular component of kinetoplast DNA of *Crithidia acanthocephali*. *Nucleic Acid Res.*, 6:3785–3804.
16. Hayduk, S. L. (1982): *Annu. Rev. Biochem. (in press)*.
17. Hoeijmakers, J. H. J., and Horst, P. (1978): RNA from the insect trypanosome *Crittidia luciliae* contains transcripts of the maxi-circle and not of the mini-circle component of kinetoplast DNA. *Biochem. Biophys. Acta*, 521:407–411.
18. Katzen, A. L., Gordon, M. C., and Blackburn, E. H. (1981): Sequence-specific fragmentation of macronuclear DNA in a holotrichous ciliate. *Cell*, 24:313–320.
19. Klaisen, C. M., and Borst, P. (1976): Sequence heterogeneity of the mini-circles of kinetoplast DNA of *Crithidia luciliae* and evidence for the presence of a component more complex than mini-circle DNA in the kinetoplast network. *Biochim. Biophys. Acta*, 407:473–478.

20. Kleisen, C. M., Weislogel, P. O., Fonck, K., and Borst, P. (1976): The structure of kinetoplast DNA: II. Characterization of a novel component of high complexity present in the kinetoplast DNA network of *Crithidia luciliae*. *Eur. J. Biochem.*, 64:153–160.
21. Leon, W., Frasch, A. G. C., Hoeijmakers, J. H. J., Fase-Fowler, F., Borst, P., Brunel, F., and Davison, J. (1980): Maxi-circles and mini-circles in kinetoplast DNA from *Trypanosoma cruzi*. *Biochem. Biophis. Acta*, 607:221–231.
22. Manning, J. E., Wolstenholme, D. R. (1976): Replication of kinetoplast DNA of *Crithidia acantocephali*. *J. Cell. Biol.*, 70:406–418.
23. Newton, B. A. (1976): Amplification of kinetoplast DNA in *Trypanosoma brucei*. In: *Biochemistry of Parasites and Host-Parasite Relationships*, edited by H. Van den Bosche, pp. 203–209. North-Holland, Amsterdam.
24. Orgel, L. E., and Crick, F. H. C. (1980): Selfish DNA: the ultimate parasite. *Nature*, 284:604–607.
25. Ritossa, F. (1973): Crossing-over between X and Y chromosomes during ribosomal DNA magnification in *Drosophila melanogaster*. *Proc. Natl. Acad. Sci. USA*, 70:1950–1954.
26. Riou, G., and Saucier, J. (1979): Characterization of the molecular components in kinetoplast-mitochondrial DNA of *Trypanosoma equiperdum*. *J. Cell. Biol.*, 22:248–263.
27. Riou, G. F., Belnat, P., and Bernard, I. (1980): Complete loss of kinetoplast DNA sequences induced by ethidium bromide or by acriflavine in *Trypanosoma equiperdum*. *J. Bioi. Chem.*, 255:5141–5144.
28. Roozen, K. J., Fenwick, R. G., and Curtiss, R. III (1971): Syntesis of ribonucleic acid and protein in plasmid-containing minicells of *E. coli* R-12. *J. Bacteriol.*, 107:21–33.
29. Simpson, A. M., and Simpson, L. (1974): Labeling of *Crithidia fasciculata* DNA with ^{3}H thymidine. *J. Protozool.*, 21:379–382.
30. Simpson, A. M., and Simpson, L. (1976): Pulse-labeling of kinetoplast DNA: localization of two sites of synthesis within the networks and kinetics of labeling of closed minicircles. *J. Protozool.*, 23:583–587.
31. Simpson, L., and Simpson, A. M. (1978): Kinetoplast RNA of *Leishmania tarentolae*. *Cell*, 14:169–178.
32. Simpson, L., Simpson, A. M., Kidane, G., Livingston, L., and Spithill, T. W. (1980): The kinetoplast DNA of the hemoflagellate protozoa. *Am. J. Trop. Med. Hyg. (Suppl. 5)*, 29:1053–1063.
33. Steinert, M., and Van Assel, S. (1980): Sequence heterogeneity in kinetoplast DNA: reassociation kinetics. *Plasmid*, 3:7–17.
34. Steinert, M., Van Assel, S., Borst, P., and Newton, B. A. (1976*a*): Evolution of kinetoplast DNA. In: *The Genetic Function of Mitochondrial DNA*, edited by C. Saccone and A. M. Kroon, pp. 71–81. North-Holland, Amsterdam.
35. Steinert, M., Van Assel, S., and Steinert, G. (1976*b*): Mini-circular and non-mini-circular components of kinetoplast DNA. In: *Biochemistry of Parasites and Host-Parasite Relationship*, edited by H. Van den Bossche, pp. 193–202. North-Holland, Amsterdam.
36. Tobler, H. (1975): DNA amplification. In: *Biochemistry of Animal Development, Vol. 3: Molecular Aspects of Animal Development*, pp. 91–143. Academic Press, London.

Molecular Biology of Parasites, edited by
J. Guardiola, L. Luzzatto, and W. Trager.
Raven Press, New York © 1983.

Investigation of the DNA of the Human Malaria Parasite *Plasmodium falciparum* by *in vitro* Cloning into Phage λ

N. Bone*, T. Gibson*, M. Goman*, J. E. Hyde*, G. W. Langsley*, J. G. Scaife*, D. Walliker†, N. K. Yankofsky*, and J. W. Zolg*

Departments of Molecular Biology and Genetics†, University of Edinburgh, Edinburgh EH9, Scotland*

The malaria parasite, *Plasmodium falciparum*, has been extensively studied biochemically (Homewood, 1978; Sherman, 1979), but much remains to be learned about this organism at the molecular level. In particular, studies on the molecular genetics of the parasite could provide information of clinical importance.

To this end we have isolated the RNA and DNA from erythrocytic forms of the parasite. The latter has been cloned into phage λ to produce a library of fragments in which most of the genome is represented.

We present here studies on the cloned DNA and describe the characterization of different cloned fragments containing repetitive DNA and ribosomal genes, respectively.

ISOLATION AND CHARACTERIZATION OF DNA FROM PARASITES

In preparing whole-cell DNA from malaria parasites, we were aware that they contain nucleases of various kinds and thus designed our extraction procedures to protect the integrity of the DNA (Miller and Ilan, 1978; Banyal, et al., 1981). Two different extraction methods prove to be equally successful in yielding largely double-stranded DNA of high molecular weight. Cultured parasites (Trager and Jensen, 1977; Zolg et al., 1981) at all stages of the erythrocytic cycle (2 to 8 $\times$ 10^9 total) are freed from host cells by gentle saponin lysis. Briefly, the DNA is extracted from the parasites by detergent lysis (sodium lauryl sarcosine) and either incubated with excess proteinase K (Firtel and Bonner, 1972) and banded in a CsCl/EtBr equilibrium gradient or, alternatively, extracted with phenol in the presence of nuclease inhibitors (Kleisen et al., 1975) and banded in CsCl without EtBr (Fig. 1).

By either method the DNA appears as a single band, and the RNA collects as a pellet at the bottom of the tube. Control experiments with uninfected red blood cells gave no detectable nucleic acid. In addition, the G + C content of the DNA

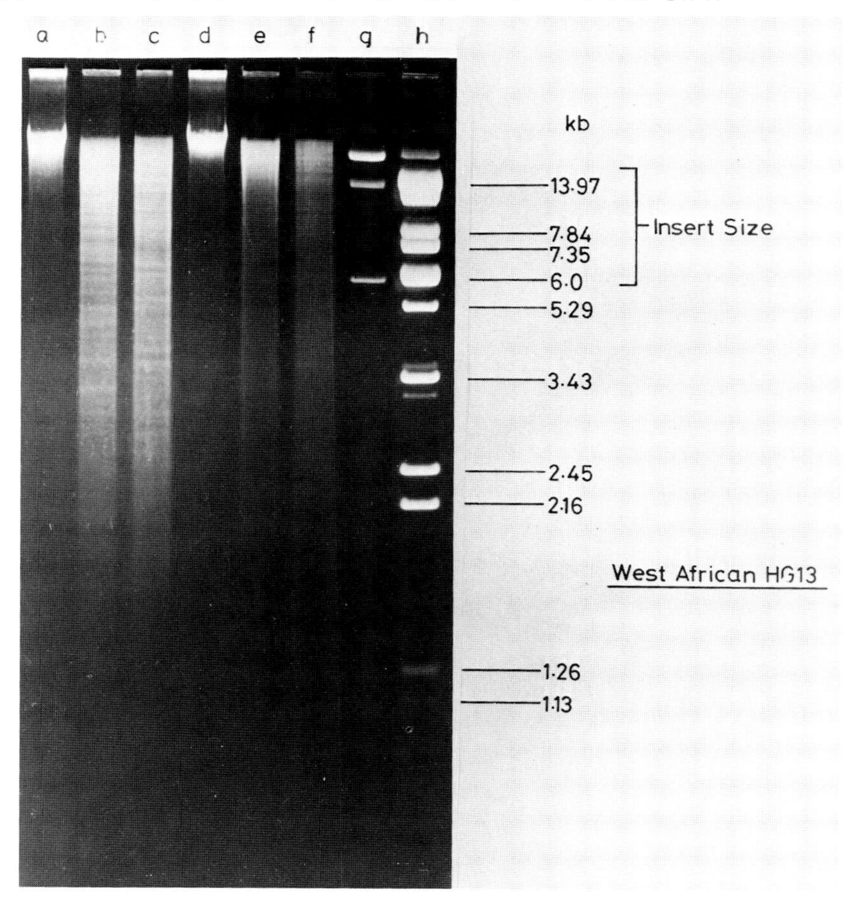

FIG. 1. DNA extracted from two independent cultures of *P. falciparum*. Parasites (about 5×10^9) grown in human red blood cells *in vitro* (Trager and Jensen, 1976) are released by adding 10^{-2} volume of 10% Saponin, incubated (0°C, 3–5 min), spun (4,500 rpm, 4°C, 10 min), and washed with TE buffer (Tris-HCl, 10^{-2} *M*, pH 8.0). Parasites in 4 ml of TE buffer are lyzed with sodium lauryl sarcosine (final concentration, 4%) in the presence of proteinase K (1 mg/ml) for 1 hr at 37°C. The total product is run on a CsCl gradient (ρ = 1.55 g/cm^3; ethidium bromide, 500 μg/ml; 48 hr; 38,000 rpm; 18°C in a Beckmann 50 Ti rotor). The DNA forms a single band that is collected from the side of the tube and run over Dowex 50W-Y8 to remove the ethidium bromide, dialyzed in TE buffer (three changes), ethanol-precipitated, and stored as a dry pellet (4°C). The parasites for the DNA in this figure were kindly supplied by Dr. G. Butcher. **a–c:** DNA from culture 1: undigested (a), digested with *Hin* dIII (b), and digested with *Eco* RI (c). **d–f:** DNA from culture 2: undigested (d), digested with *Hin* dIII (e), and digested with *Eco* RI (f). **g:** λNM788 digested with *Hin* dIII, showing the phage arms (top two bands) and the insert. H: Size markers.

we obtain is quite different from the host's (see below), confirming that it originates in the parasites.

The DNA prepared in this way has been examined in the electron microscope. Most of the molecules are double-stranded, without extensive single-strand gaps. The majority are between 50 and 150 kilobase pairs (kb) long, but smaller molecules

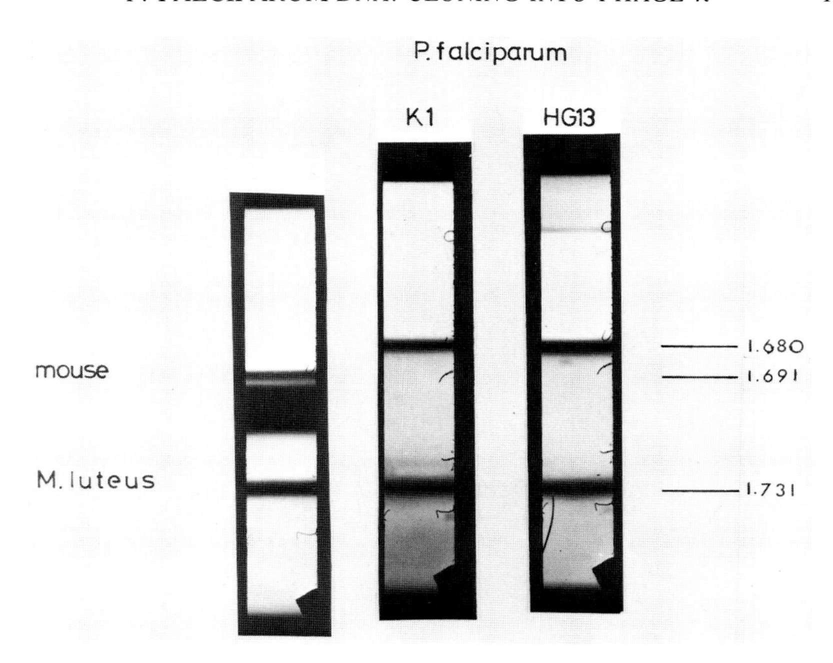

FIG. 2. *P. falciparum* DNA. Density measurements by equilibrium gradient centrifugation. DNA (2 μg) in TE buffer; initial ρ = 1.600 g/cm³; 44,000 rpm; 20 hr; 25°C. Markers: mouse satellite DNA (21.691) and *Micrococcus luteus* DNA (ρ = 1.731).

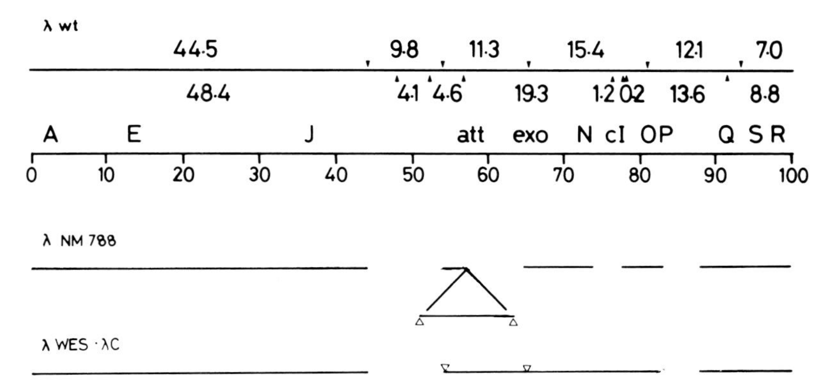

FIG. 3. Vectors for DNA cloning. The top two lines show the parent bacteriophage λ, its sites for *Hin* dIII, and *Eco* RI (upper line), and a summary genetic map of the phage (lower line). The distances between restriction sites are in kilobases. λNM788 and λWESλC. The maps are drawn to scale. The gaps represent deletions of DNA from the parent phage. The insert in λNM788 (drawn below its point of insertion) is a *Hin* dIII fragment from *E. coli* carrying *trpE.* Arrows above the line are *Eco* RI sites; those below are *Hin* dIII sites. Open arrows signify those used for cloning.

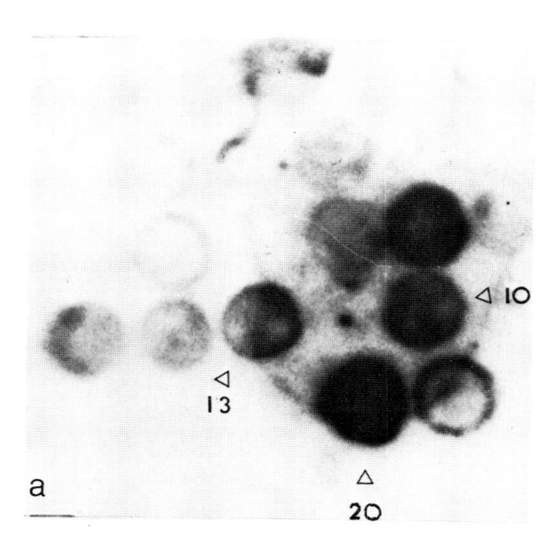

FIG. 4. Clones of *P. falciparum* repetitive DNA in λNM788. **A:** Plaques screened as putative repetitive DNA phage were picked and spotted on a lawn of indicator bacteria to make patches of lysis. The patches were transferred to filters and hybridized against nick-translated parasite DNA (Benton and Davis, 1977). The arrows indicate those patches that later proved to have repetitive DNA. **B:** Southern blots of *P. falciparum* DNA against λNM788Pf 10, 13, and 20. Panel I (probed with λNM788pf10): track 1, size markers; 2, *Hin* dIII-cut K1 DNA; *Hin* dIII-cut K28 DNA; 4, λNM788pf10. Panel II (probed with λNM788pf13): track 1, *Hin* dIII-cut K1 DNA; 2, *Hin* dIII-cut K28 DNA; 3, λNM788pf13. Panel III: track 1, *Hin* dIII-cut K1 DNA; 2, *Hin* dIII-cut K28 DNA, track 3: λNM788pf20.

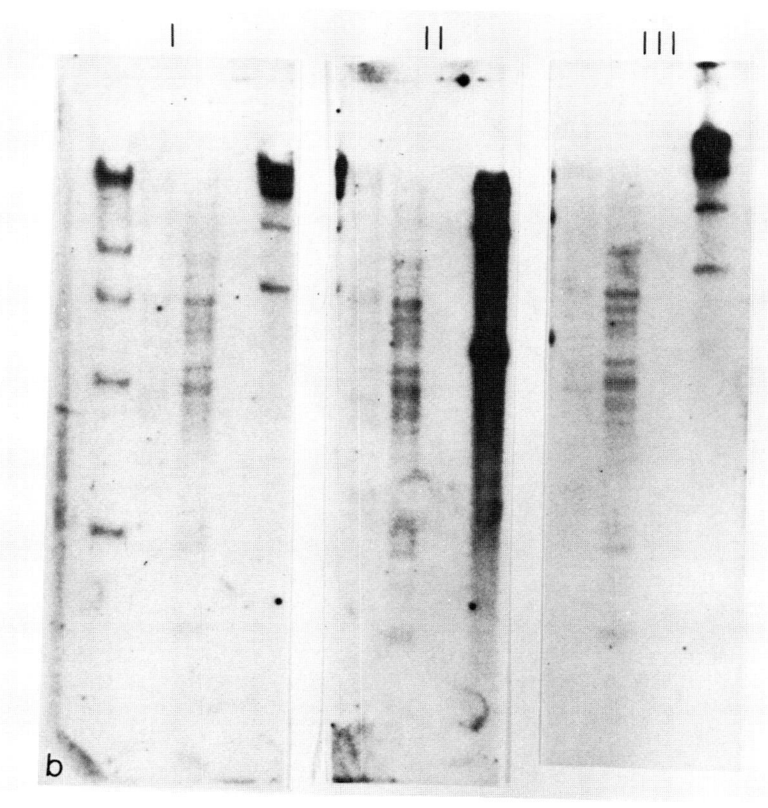

TABLE 1. *Libraries of* P. falciparum *DNA*

Vector[a]	DNA source[b]	Nature of fragments	Number of independent plaques[c]	Types of clone recovered
λNM788	HG13	*Hin* dIII	20,285	(a) Repetitive DNA λNM788Pf10 λNM788Pf13 λNM788Pf20 (b) rRNA genes λNM788Pf2 λNM788Pf5
λgt*WES*.λC	K1	*Eco* RI	21,000	

[a]Phages were grown as liquid lysates on TGL 70 *met sup*E *sup*F *hsd*R *ton*A, treated with DNAse, pelleted in the centrifuge (19,000 rpm; 3 hr), banded in CsCl gradients, and dialyzed against Tris (10 m*M*) EDTA (1 m*M*) buffer, pH 8.2. DNA was phenol-extracted, digested with the appropriate restriction endonuclease, and sedimented on neutral sucrose gradients to separate the insert.

[b]HG13 parasites are of West African origin kindly supplied by Dr. G. Butcher. K1 comes from Thailand. All parasites were grown by the candle-jar method (Jensen and Trager, 1977) and harvested at 4–10% parasitemia.

[c]Recombinant phage were propagated on NB16 *met sup*E *sup*F *hsd*R *lac ton*A *rec*A to guard against loss of insert DNA by recombination. This strain, like TGL70, lacks the *E. coli*-K12-specific restriction system but retains the modification system so that progeny phage can be subsequently grown on all strains of *E. coli* K12.

are present. Exhaustive searches of several preparations have revealed no circular molecules.

The same DNA has been electrophoresed on agarose gels (Fig. 1). The bulk of the DNA migrates as a broad band in the range of 50 to 150 kb, confirming our electron-microscopic observations.

The purified DNA on a dye-free CsCl gradient equilibrates in a single band at a density of 1.68 g/cm^3, equivalent to an average G + C content of 19% (Fig. 2). An earlier study (Gutteridge et al., 1971) used the DNA of *P. falciparum* grown in aotus monkeys. This DNA resolved into two components on CsCl gradients with guanidine plus cytosine (GC) contents of 37 and 19%. There are several plausible explanations for this difference, notably the possibility that host DNA (GC = 37%) was not entirely excluded from the earlier monkey-grown preparation. White blood cell contamination is not a problem in the culture system we use. Otherwise, there could be wide variation in the GC content within the genome, which might give two species in low-molecular-weight material.

The two methods outlined above gave 0.01 to 0.02 pg DNA/parasite nucleus, quantities comparable to previously reported yields (Dore et al., 1980).

The DNA can be cut by restriction enzymes. The fragments produced generate a distinct pattern on agarose gels (Fig. 1). This feature of the DNA could be explained if *P. falciparum* DNA, like that of *P. berghei* (Dore et al., 1980), contains

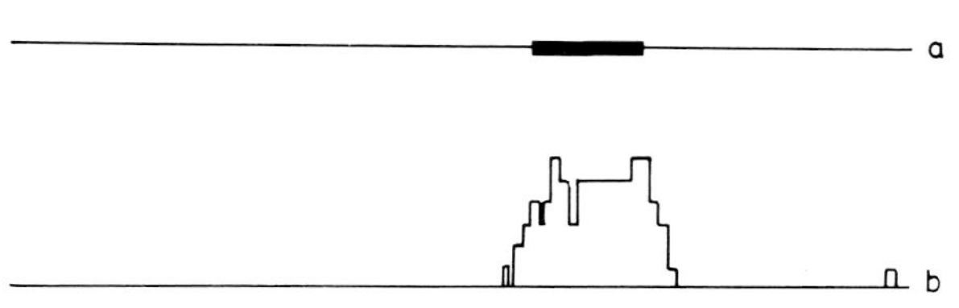

FIG. 5. Partial denaturation of λNM788pf13. **a:** Representation of the phage, showing the *P. falciparum* insert (thick line). **b:** Histogram of the fraction of molecules denatured at each position. Essentially all the molecules were denatured in the region of the insert but not in the phage DNA on each side. The site on the right arm has the highest AT content of phage λ.

a repetitive fraction. Qualitative support for this conclusion comes from electron-microscopic observations of whole *P. falciparum* DNA, denatured and reannealed to allow repetitive sequences to reform duplex molecules. Under conditions optimal for the formation of heteroduplexes between DNA molecules of different λ phages, double-stranded molecules do emerge from total *P. falciparum* DNA. We estimate that approximately 20% of the denatured parasite DNA has reannealed and is thus repetitive. This conclusion is confirmed by analysis of *P. falciparum* DNA fragments cloned into bacteriophage λ (see below).

INSERTION OF RESTRICTION ENDONUCLEASE FRAGMENTS OF *P. FALCIPARUM* DNA INTO BACTEROPHAGE λ

Much information about the genetic organization of *P. falciparum* can accrue from cloning DNA fragments into a suitable prokaryotic vector. We have used derivatives of phage λ, which allow the insertion of fragments created by restriction endonucleases. One of those λNM788 (Murray et al., 1977) allows *Hin* dIII fragments to be cloned. It has two *Hin* dIII sites located at the junctions of the two arms of the phage and a segment of the *E. coli trp* operon inserted between them (Fig. 3). The *trp* insert can be replaced *in vitro* by *Hin* dIII-cleaved foreign DNA of similar size to create a viable recombinant stably propagating the new fragment. A similar system, λ*gt WES* (Leder et al., 1977) has been used to clone *Eco* RI fragments (Fig. 3).

We have made a large collection of such phages constituting a library of independent *P. falciparum* DNA fragments, each inserted into a phage vector molecule. Precautions were taken that should ensure that more than 90% of the parasite genome is represented in the collection. The phage library provides a simple way of propagating the genome. At the same time, the individual phages, each containing a single fragment when isolated and studied in detail, can tell us about the organization of the genome.

The manufacture of the phage library has four basic steps:

1. purification of the vector and parasite DNA,
2. *in vitro* recombination between them,
3. packaging recombinant molecules *in vitro* into phage coats, and
4. growth of the resulting phages.

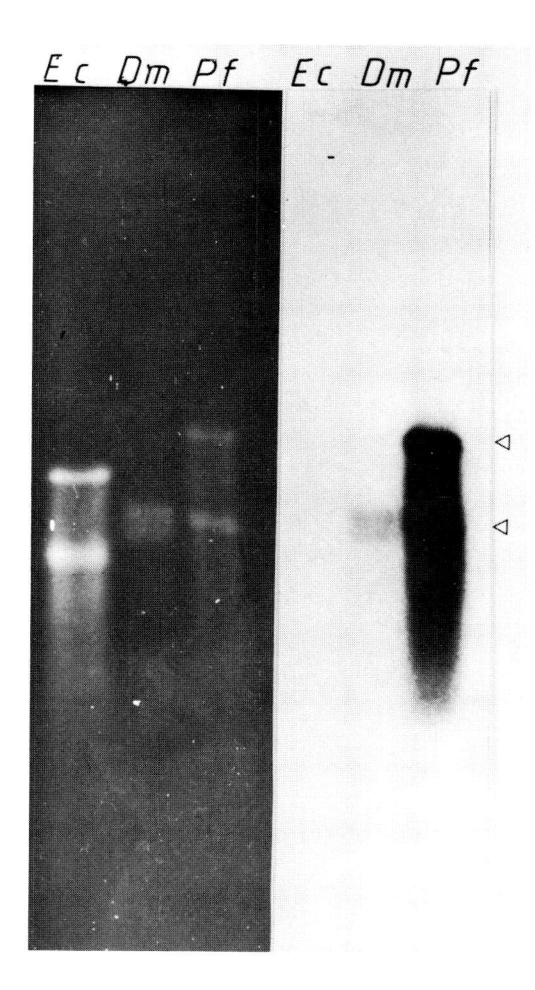

FIG. 6. A λNM788Pf clone containing ribosomal RNA genes. Left panel (dark background) shows ribosomal RNA run on agarose gels from *E. coli* (Ec), *Drosophila melanogaster* (Dm), and *P. falciparum* (Pf). The RNA was transferred to nitrocellulose paper (Langsley et al., *unpublished*) and hybridized to nick-translated NM788Pf2. Note that *E. coli* rRNA does not hybridize, *D. melanogaster* rRNA hybridizes weakly, and *P. falciparum* rRNA hybridizes very strongly. The arrows show the large and small rRNA species of *P. falciparum*. The fact that both hybridize to λNM788Pf2 shows that this phage has at least part of both genes.

After restriction-endonuclease digestion, the vector DNA contains the *E. coli trp* insert, which would compete with *P. falciparum* DNA in the recombination reaction. It is therefore removed by sucrose gradient centrifugation of the digest, which resolves the phage arms (large) from the insert (small) and allows the former to be recovered pure (Maniatis et al., 1978).

The distribution of the restriction endonuclease target sites in the parasite genome is not regular. As a result, the restriction fragments obtained vary in size (Fig. 1). Not all of these can fit into a viable phage. The capacity of the phage head is finite, imposing an upper limit on the size of the fragment. There is also a lower limit determined either by the packaging machinery or by the inability of underpacked phages to inject their DNA into a bacterium. These limits have statistical implications considered below. They also mean that the best yield of viable recombinant phage can be obtained by selecting fragments within the acceptable size range for the recombination reaction. This is done by sucrose gradient centrifugation.

The separated phage arms and sized *P. falciparum* DNA fragments are joined *in vitro* by annealing their endonuclease-specific, single-stranded termini and ligated. The recombinant molecules are then mixed with extracts of bacteria, making large quantities of the phage packaging machinery. The recombinant molecules are packed and yield complete phage particles able to infect and grow in bacteria (Hohn and Murray, 1977; Scalenghe et al., 1981).

The phages are plated on lawns of indicator bacteria and form plaques, which are either collected as a mixture or picked individually into wells of microtiter plates and stored for future use.

It is evident that this technique will select against certain segments of the genome. Any segment between *Hin* dIII sites too near or too distant to give a clonable fragment will be lost. The fraction of fragments expected to be in this category can be estimated if we assume that *Hin* dIII sites are randomly distributed in the *P. falciparum* genome. It is ≃15%. For this reason, we made the second library containing fragments cut by another enzyme, *Eco* RI. The vector in this case was (Table 1) λ*gt.WES* (Leder et al., 1977) (Fig. 3).

Each of the libraries (Table 1) contains more than 20,000 recombinants, a quantity that ensures ($P = 0.99$) that all packagable fragments of the genome will be represented (Clarke and Carbon, 1976).

ISOLATION AND CHARACTERIZATION OF λNM788 *P. falciparum* PHAGES CONTAINING REPETITIVE DNA

Reannealing studies show that as much as 18% of the DNA from the rodent parasite *P. berghei* is repetitive (Dore et al., 1980). Likewise, we have outlined evidence above suggesting that part of the *P. falciparum* genome is repetitive. It was of interest to discover whether we could detect clones in the libraries containing repetitive DNA.

We reasoned that such clones should be readily detected by screening for phage plaques, whose DNA hybridize to whole *P. falciparum* DNA labeled with ^{32}P-containing dCTP by nick-translation. A clone containing repetitive DNA, unlike those with unique sequences, should readily find homologous, labeled molecules in the probe and become heavily labeled itself.

The library was plated to form single plaques, which were transferred to nitrocellulose filters and treated to release, denature, and fix their DNA to the matrix (Benton and Davis, 1977). When the filters were hybridized to the probe DNA and exposed to x-ray film, we did find a proportion (~20%) of phage plaques that had become labeled. Three of these were selected for further study: λNM788Pf10, λNM788Pf13, and λNM788Pf20 (Fig. 4).

Digestion with *Hin* dIII (Fig. 4) excises the inserts and shows that each is different in size from the others and from that of the vector (7.05, 5.0, and 7.1 kb for λNM788Pf10, 13, and 20, respectively).

Hybridization of ^{32}P-labeled DNA from the three phages to whole, *Hin* dIII-cut *P. falciparum* DNA shows two features (Fig. 4). The whole track has hybridized to the probe to a low degree, showing that each phage contains sequences present in very many parts of the genome. In addition, a number of discrete bands are heavily labeled, strongly suggesting that the phages also contain DNA sequences that occur in specific fragments.

Interestingly, our hybridization studies indicate that the specific fragments vary from one *P. falciparum* isolate to another. We are currently investigating whether this property might be used to distinguish between isolates.

Finally, partial denaturation studies on λNM788Pf13 show that its insert is, as expected, adenine plus thymine (AT)-rich. Detailed maps exist for phage λ showing those regions that, because they have higher AT content, denature under mildly alkaline conditions to form bubbles in the viral DNA visible in the electron microscope (Fig. 5). Moreover, direct estimates are available (P. Highton, unpublished) for the AT content of each of these regions. When denatured under conditions that barely affect the vector DNA, the *P. falciparum* insert λNM788Pf13 is extensively denatured (Fig. 5), although the conditions used are so mild that region in the right arm of the vector with sequences close to 70% AT does not denature. It is reasonable to conclude that the insert DNA has more than 70% AT.

IDENTIFICATION OF λNM788Pf PHAGES CONTAINING RIBOSOMAL RNA GENES

The ribosomal RNA (rRNA) of *P. falciparum* has been purified and characterized (Hyde et al., 1981). Its major components are a large species of molecular weight $(1.49 \pm 0.09) \times 10^6$ and a small species of $(0.78 \pm 0.02) \times 10^6$ present in equimolar quantities. The overall base composition of these species is 40% GC.

We have been able to use the cloned rRNA genes of *Drosophila melanogaster* to extract from our *Hin* dIII library the rRNA genes of *P. falciparum*. The *Drosophila* clone, pDm103 (Glover and Hogness, 1977) hybridizes both to specific fragments in total digested *P. falciparum* DNA and to purified *P. falciparum* rRNA (data not shown).

One of the phages containing *P. falciparum* rRNA genes is λNM788Pf2. It contains an insert 8 kb long (data not shown). As expected, the insert hybridizes to pDM103. In addition, the new phage contains sequences that hybridize to both large and small rRNA species from *P. falciparum* (Fig. 6). We conclude that this clone contains at least a part of both genes for the major rRNA species. The analysis of this and related clones is under way to investigate the organization of these genes in the parasite.

ACKNOWLEDGMENTS

We pay tribute to Professor G. Beale for his support and encouragement during the establishment of this project. We thank Ian Purdom for performing the CsCl gradient analysis of *P. falciparum* DNA and Dr. G. Butcher for providing the Hg13 isolate of *P. falciparum*. One of us (J.W.Z.) is a fellow of the Cusanuswerk Bischöfliche Stüdienforderung, Bonn. N.K.Y. is a British Council Scholar. This work was supported by the Medical Research Council, United Kingdom.

REFERENCES

1. Banyal, H. S., Pandey, V. C., and Dutta, G. P. (1981): Ribonucleases and deoxyribonucleases in *Plasmodium knowlesi*. *Ind. J. Exp. Biol.*, 19:173–175.
2. Benton, W. D., and Davis, R. W. (1977): Screening λgt recombinant clones by hybridisation to single plaques *in situ*. *Science*, 196:180–182.
3. Clarke, L., and Carbon, J. (1976): A colony bank containing synthetic col El hybrid plasmids representative of the entire *E. coli* genome. *Cell*, 9:91–99.
4. Dore, E., Birago, C., Frontali, C., and Battaglia, P. A. (1980): Kinetic complexity and repetitivity of *Plasmodium berghei* DNA. *Mol. Biochem. Parasitol.*, 1:199–208.
5. Firtel, R. A., and Bouner, J. (1972): Characterisation of the genome of the cellular slime mold *Dictyostelium discoideum*. *J. Mol. Biol.*, 66:339–361.
6. Glover, D. M., and Hogness, D. (1977): A novel arrangement of the I8S and 28S sequences in a repeating unit of *D. melanogaster* rDNA. *Cell*, 10:167–176.
7. Gutteridge, W. E., Trigg, P. I., and Williamson, D. H. (1969): Base compositions of DNA from some malarial parasites. *Nature*, 224:1210–1211.
8. Hohn, B., and Murray, K. (1977): Packaging recombinant DNA molecules into bacteriophage particles *in vitro*. *Proc. Natl. Acad. Sci. USA*, 74:3259–3263.
9. Homewood, C. A. (1978): In: *Rodent Malaria*, edited by R. Killick-Kendrick and W. Peters, pp. 169–211. Academic Press, London.
10. Hyde, J. E., Zolg, J. W., and Scaife, J. G. (1981): Isolation and characterisation of ribosomal RNA from the human malaria parasite *Plasmodium falciparum*. *Mol. Biochem. Parasitol.*, 4:283–290.
11. Kleisen, C. M., Borst, P., and Weijers, P. J. (1975): The structure of kinetoplast DNA. I. Properties of the intact multicircular complex from *Critihidia luciliae*. *Biochim. Biophys. Acta*, 390:155–167.
12. Leder, P., Tiemeier, D., and Enquist, L. (1977): EK2 derivatives of bacteriophage λ useful in the cloning of DNA from higher organisms. The λgtWES system. *Science*, 196:175–177.
13. Maniatis, T., Hardison, R., Lacy, E., Lauer, J., O'Connell, C., Quon, D., Gek kee Sin, and Efstratiadis, S. (1978): The isolation of structural genes from libraries of eukaryotic DNA. *Cell*, 15:687–680.
14. Miller, F. W., and Ilan, J. (1978): The ribosomes of *Plasmodium berghei*: isolation and ribosomal ribonucleic acid synthesis. *Parasitology*, 77:345–346.
15. Murray, N. E., Brammar, W. J., and Murray, K. (1977): Lambdoid phages that simplify the recovery of *in vitro* recombinants. *Mol. Gen. Genet.*, 150:53–61.
16. Scalenghe, F., Turco, E., Edström, J. E., Pirrotta, V., and Melli, M. (1981): Microdissection and cloning of DNA from a specific region of *Drosophila melanogaster* polytene chromosomes. *Chromosoma (Berl.)*, 82:205–216.
17. Trager, W., and Jensen, J. B. (1976): Human malaria parasites in continuous culture. *Science*, 193:673–675.
18. Zolg, J. W., McLeod, A. J., Dickson, I. H., and Scaife, J. G. (1981): *Plasmodium falciparum*: modification of the *in vitro* culture conditions improving parasite yields. *J. Parasitol. (in press)*.

Molecular Biology of Parasites, edited by
J. Guardiola, L. Luzzatto, and W. Trager.
Raven Press, New York © 1983.

Mechanisms of Disease Resistance

R.K.S. Wood

Imperial College, London SW7 2BB, United Kingdom

This chapter is about diseases of higher plants caused by parasitic bacteria and fungi. First, a few general points about these diseases. The relations between plants and parasites are almost always specific. A particular parasite causes disease in a small number of plant species, usually closely related ones. A particular plant is susceptible only to a small number of the very many bacteria and fungi that can cause disease in higher plants. A few parasites are, however, much less specialized. They are not included in this chapter, although some are important pathogens of crop plants.

Next, there are the nutritional relations between plants and their parasites. What for convenience will be called a facultative parasite grows well on simple media, and it can be safely presumed that almost always it will be able to grow on dead material both of plants it can and of the many other plants that it cannot parasitize. In contrast, obligate parasites, with a few exceptions, have not yet been grown in culture, and they grow little if at all on dead material of plants they can parasitize or of other plants. For sustained growth, obligate parasites must be closely associated with living cells.

The last and obvious point, although often neglected, is that to cause disease parasites almost always must grow in plants. Disease resistance largely depends, therefore, on properties of resistant plants that prevent the growth in them of potential parasites. There are, of course, a few exceptions to these generalities; no doubt some will be referred to in later chapters.

Properties that make plants resistant are of two classes. First are those that are preformed in the sense that they are found in healthy plants and are therefore independent of the parasite. Thus, a tissue may be resistant because the walls of its cells are lignified and are degraded only slowly, if at all, by a parasite. The second are induced in that they develop in response to challenge by the parasite, as, for example, when the walls of the cells of healthy tissue are not lignified but become so in response to infection.

For plant diseases generally there is abundant evidence that induced resistance is by far the more important. It is therefore the main subject of this paper.

A large part of research on induced resistance is and has been about diseases of crop plants caused by highly specialized parasites. Almost invariably in such diseases the parasite occurs in races that are identified by their different capacities to cause disease in different cultivars of the plant. In a good many diseases, genetic analyses have shown that for a certain, important type of resistance, single genes for resistance in the plant are matched by single genes for low virulence (for convenience, avirulence) in the parasite, and similarly for susceptibility and virulence. In the many diseases studied from this point of view, genes for resistance are almost always dominant. Unfortunately, only in a few diseases has it been possible to establish the status of the corresponding genes in the parasite, and then avirulence too is dominant (10,14,31). Mutation of the gene for avirulence will make the parasite virulent. There can therefore be considerable selection pressure favoring such mutants in populations of parasites that are challenging plants with those genes for resistance. An interesting point emerging from recent work is that such mutants may not be so virulent as are wild types of parasites on cultivars not carrying the resistance gene. The virulent mutant does not fully overcome the resistance controlled by the single gene for resistance (25,27).

Another important form of resistance is controlled not by a single gene with a pronounced effect but by many genes (33). Now the effects of the individual genes on resistance are not known. Presumably they are small but additive. Otherwise, they may be similar to the genes with major effects, and again with matching genes for avirulence in the pathogen. There is also some speculation that they are the partially nullified ("ghost") resistance genes referred to above.

EXPRESSION OF RESISTANCE

Again referring to interactions between races of parasites and cultivars of plants and single genes with major effects, the greatest resistance occurs as the so-called hypersensitive reaction (HR), which was first described many years ago for diseases caused by the obligately parasitic rust fungi (35). It has since been described for many other types of obligate parasites and many facultative parasites, including bacteria, none of which are known as facultative parasites (24,26,32). Classically, in HR the early growth of the parasite close to, on, or in a plant cell kills this cell. A small number of adjacent cells may also be killed. The parasite then grows little if at all and is usually confined to the dead cell(s) or the immediate vicinity. The parasite may or may not remain alive; almost always this has not been determined. Although it has been usual to confine HR to reactions in which cells are killed, there is no good reason why HR should not also be used when cells are not killed. It is mistaken to assume that it is only the dead or visibly damaged cells that are important in HR. There is plenty of evidence of structural and metabolic changes in cells well away from cells visibly altered by the parasite. The most important feature of HR is perhaps not so much the appearance of reacting cells containing or close to the parasite but rather the severely restricted growth of the parasite.

At the other end of the range of interactions between races and cultivars is the fully susceptible reaction in which the parasite grows extensively and the plant is correspondingly damaged. The damage does, of course, take many forms in the many different types of disease. In particular they are usually quite different in

living cells can lead to concentrations that can kill them. The death of cells in HR would then be an irrelevant after effect, although, of course, the phytoalexins would remain the final reason for disease resistance.

The facts and ideas discussed so far suggest that in gene-for-gene systems for avirulent parasite and resistant cells, a gene-controlled determinative phase causes events that lead to an expressive phase, i.e., the synthesis and accumulation of phytoalexins that act directly against the parasite (or to other events such as lignification which prevent its growth). This is the next topic to be discussed.

RELATIONS BETWEEN DETERMINATIVE AND EXPRESSIVE PHASES

First, no one has yet convincingly and *fully* reproduced the patterns of resistance and susceptibility between races of a parasite and host cultivars by applying specific substances to the cultivars—including, even, cell-free extracts from parasites. One hopes that it will not be long before this is accomplished (2,7,22).

What then has been achieved? In the first place, extracts of dead cells of *Phaseolus vulgaris* applied to hypocotyls are quite effective in causing the phytoalexin phaseollin to accumulate (16). Substances such as those in the extracts that are active in this way are now commonly referred to as elicitors. It is postulated that cells of the hypocotyl contain a constitutive elicitor for the synthesis of phaseollin. The product of the gene for avirulence reacts with the product of the gene for resistance to kill or damage the cell. This in turn causes the release of the constitutive elicitor, which initiates the synthesis of phytoalexins in adjacent living cells. The phytoalexins diffuse into and accumulate to high concentrations in the dead or damaged cells and, indeed, may also accumulate sufficiently to kill living cells. And, of course, they accumulate to concentrations that prevent growth of the parasite.

This hypothesis rests mainly on the release of the constitutive elicitor, which certainly exists. It does not depend on the death of the plant cell, which is often mistakenly overemphasized in appraisals of disease resistance. The hypothesis also requires that cells responding to the constitutive elicitor remain alive or metabolically active long enough for the phytoalexins to accumulate to effective concentrations. Otherwise, the parasite would continue to grow and kill cells.

In the susceptible reaction in which the gene for avirulence, for resistance, or both is absent, the cells remain alive or are inappropriately damaged, so that the constitutive elicitor is not released and phytoalexins are not synthesized. This allows the parasite to grow. If it be obligate, then it will continue to grow for many days and still not kill cells. But virulent facultative parasites soon begin to do so and on an increasing scale, so that many cells are killed. The question now arises as to whether this killing of cells in susceptible reactions also releases the constitutive elicitor and, if so, whether it also initiates the synthesis of phytoalexins. They certainly do accumulate in susceptible reactions, but it is argued that a virulent parasite now kills cells too quickly to allow sufficiently rapid synthesis of phytoalexins in the adjacent living cells and therefore their accumulation, which would prevent parasitic growth.

Note again that in this argument, elicitor refers to the substance released by damaged or killed cells that initiates the synthesis of phytoalexins. There remains the question as to how the cells are killed or damaged in the first place.

Another research hypothesis holds that elicitors are substances derived from parasites, which when applied to plant tissues in low concentrations cause the accumulation of phytoalexins (20,21). Note, however, the possibility that such substances may act in the first place by damaging or killing host cells, which leads to the release of the constitutive elicitor already referred to. If, in fact, the determinative phase is controlled by the gene-for-gene system and leads to the production or release of substances that cause the expressive phase, as the synthesis of phytoalexins, it is important to determine the phase to which elicitor belongs—granted, of course, the present difficulties in separating the phases.

Turning to elicitors from parasites, there are, firstly, the β-glucans, which have been extracted by drastic treatment of the cell walls of a number of fungi (4). They certainly cause the accumulation of phytoalexins when applied to plant tissues in very low concentrations. But they do so nonspecifically, and it is therefore unlikely that they are significant primary determinants of race-specific resistance. There is, however, growing evidence for the importance of glycoproteins extracted from cell walls or from culture fluids, in spite of earlier work showing that low concentrations of certain of these glycoproteins from *Phytophthora megasperma* f. sp. *glycinea* caused an accumulation of phytoalexins that was only nonspecific, both as regards races and cultivars. On the basis of other very similar work, it has been claimed that extracts from cell walls of races of this parasite containing glycoproteins do cause differential accumulation of phytoalexins in cultivars in a pattern similar to that for resistance and susceptibility. This activity is lost after treatment with periodate but not pronase. It is not yet known whether the carbohydrate serving as the elicitor is part of a glycoprotein or is noncovalently associated with protein (22).

Still other similar work has not confirmed this specificity but has shown that extracellular glycoproteins from avirulent races applied to soybean plants protect against later infection by virulent races (34). It has also been found that there are differences in the carbohydrates of extracellular invertase gylcoprotein from three races of the parasite (40). Invertase glycoproteins can also be extracted from cell walls, and those from two races of the parasite contained different carbohydrates (36). Extracellular fractions containing glycoprotein from races of *Colletotrichum lindemuthianum* also differed in causing accumulation of phytoalexins in ways corresponding to their parasitism. Soluble fractions from cells of *Pseudomonas glycinea* behave similarly (2,7).

This research certainly indicates an important role for carbohydrate, probably as glycoprotein in the determinate phase of race-specific resistance. A further development is the recent finding that enzymes from plant tissues can release substances from the cell walls of fungal parasites that cause accumulation of phytoalexins when applied in low concentrations to similar plant tissues. Thus, a partially purified enzyme from soybean release from the cell walls of *P. megasperma* f. sp. *glycinea* high-molecular-weight carbohydrates containing glucose and mannose with the same specificity for causing accumulation of phytoalexins as the glycoproteins already referred to and the races of the parasite (ref. 39; and M. Yoshikawa and N.T. Keen, *personal communication*, 1981). This double-induction of synthesis and accumulation of phytoalexins suggests a role for the nonspecific elicitors that also have been obtained from parasites. Other recent work suggests that carbohydrates released from plant cells early in infection trigger the synthesis of phytoalexins.

Whatever the nature of the substances from parasites that control race-specific resistance, there must be corresponding reactants in the plant (8,30). But so far they have been little studied. Comparisons with other systems suggest that they would be in plasma membranes, and there is some evidence that they are, probably, proteins or glycoproteins. Otherwise, they must be in cell walls, and for certain bacteria it appears likely that cell wall lectins have a role in cell damage and death, although the relation between HR and decreased growth of the parasite has not been established (29).

In summary, therefore, the picture that emerges is as follows (20). Genes for avirulence in races code for the production of specific carbohydrates, probably occurring as glycoproteins and with synthesis catalyzed by specific glycosyl transferases, as predicted by Albersheim and Anderson (1). The carbohydrates may be surface-bound or extracellular. They combine with receptors in cell walls or in plasma membranes, synthesis being controlled by products of matching genes for resistance in the plant. Then, in some unknown manner, this leads to information that is transferred to the cell nucleus of the reacting cell, and probably of other cells too. There follows DNA transcription and RNA translation, which end in the synthesis of the enzymes involved in the production of phytoalexins. This model is based primarily on race-cultivar-specific interactions in which dominant genes for avirulence are matched by dominant genes for resistance, which result in major effects recognized as HR: very restricted growth of the parasite, accumulation of phytoalexins, or other mechanisms of resistance. There are, however, many other features of resistance that the model must accommodate (20). Thus, genes for virulence, not for avirulence, in the parasite may be dominant, and genes for resistance in the plant may be recessive. Can the model cope with so-called stabilizing selection in which races with superfluous genes for virulence are less fit in competition with races with fewer such genes? Will it explain reactions in the plant that are intermediate between HR and high susceptibility and responses in diseases caused by rust fungi called mesothetic which are quite different in different parts of the same plant? And can it explain race-nonspecific resistance—which is controlled by many genes, each with minor effects? We seem to be a long way from answering most if not all of these questions in biochemical terms.

OBJECTIONS TO THE ABOVE MODEL

Ellingboe (12) has objected on genetic grounds to the model described above based on single, dominant, matching genes for avirulence and resistance. First, there are the proposed reactions between glycoproteins or between carbohydrate and a glycoprotein. For the glycoprotein, one enzyme would code for the protein and one or more enzymes would code for the carbohydrate, depending on its complexity. If specificity depends on the intact molecule, then mutations of two or more genes could result in loss of specificity. But this would not conform to the known gene-for-gene relationship. Similar arguments apply to the expressive phase, which depends on the synthesis of phytoalexins; this again would depend on a number of enzymes and, therefore, on a number of genes—any of which could mutate to disrupt the synthesis. Again, the ratios for resistance to susceptibility would be different from those that are observed. Ellingboe points out that the characteristic ratio would be expected if the products of the avirulence and resistance genes combined to give another product that directly prevented growth of the parasite.

INDUCTION OF SUSCEPTIBILITY

Most earlier research on race-specific resistance has emphasized the induction of resistance as the determining mechanism. If resistance is not induced, then the parasite grows and the reaction is susceptible. But there is also the view and the evidence that the specific reactions are those that determine susceptibility not resistance (9,19). Thus, certain high-molecular-weight glucans can be extracted from races of the *Phytophthora infestans*, which, as in other diseases, nonspecifically kill and damage cells and cause accumulation of phytoalexins in cultivars of potato. But other glucans of low molecular weight can also be extracted from the same races, and these suppress the necrosis and the synthesis of the phytoalexins that would be caused by the nonspecific glucans (11). A suppressor is most effective in the cultivar susceptible to the race from which the suppressor was obtained (15). What would seem to be high concentrations ($\sim$1 to 10 mg ml^{-1}) of these glucans are required for suppression compared with the much lower concentrations of the high-molecular-weight glucans that act nonspecifically. Nevertheless, the results could imply that all races of a parasite will induce resistance, presumably in the same way, in all cultivars of a plant and that to cause disease this general response, or those that lead from it, must be suppressed. The suppression itself may be specific, as suggested by the evidence given above, or it could be nonspecific in that it would make the tissue susceptible not only to different races of the parasite, but also to parasites of other plants.

Further evidence for induction of susceptibility is that successful establishment of certain obligate parasites may make adjacent uninfected cells susceptible to related parasites that do not normally parasitize the tissue (28). Exudates from infection hyphae of rust fungi and from infected leaves can also induce susceptibility, and preliminary evidence suggests that the active substances are of relatively low molecular weight (<5,000) (17,18). Effects caused by virulent parasites seem to be limited to comparatively few cells around those originally infected.

At present, it is difficult to assess the significance of these and a few other groups of experiments along similar lines. Although there can be little doubt about the phenomenon itself, there are some difficulties in assessing its general significance. Apart from those already mentioned, there is the genetic evidence that in most diseases, resistance—not susceptibility—is dominant-gene-controlled and that in certain diseases, including some caused by obligate parasites, avirulence—not virulence—also is dominant-gene-controlled. A further point is that within populations of cultivars challenged by populations of races of parasites, it is the change from avirulence to virulence that is the rule, and for certain important crop diseases all too common. This loss of gene function would be expected. The gain in gene function required for induction of susceptibility or suppression of resistance is less likely.

HOST-SELECTIVE TOXINS

There are a few but an increasing number of diseases in which mechanisms of resistance are claimed to be quite different from those described above (38). A parasite causes disease in some cultivars of a plant. In culture it produces a substance that at low concentrations kills cells of susceptible cultivars. Very much higher concentrations must be applied to kill cells of resistant cultivars or of other plants resistant to the parasite. Such substances are called host-selective (or specific) toxins. They have now been obtained from a number of fungal parasites, and some have been identified: they are active in low, sometimes remarkably low, concentrations; there is a great difference, sometimes very great, in their toxicity to resistant and susceptible cultivars; and their specificity is usually very pronounced. In certain diseases it is known that susceptibility to the toxin and to the parasite is dominant-gene-controlled, as is production of the toxin. Thus, susceptibility corresponds to the matching of a gene controlling the production of the toxin and a gene controlling the formation of the receptor in the plant. Lack of either gene or both genes will give a resistant reaction. These relations are the reverse of those for induced resistance.

In this group of diseases it is assumed that susceptibility depends on the killing of plant cells by the host-selective toxin. The dead cells then produce substances for growth of the parasite. If plant cells are not killed by the toxin, then the substrates for growth of the parasite do not exist, and the plant is resistant.

There are a number of difficulties in this simple, satisfying, and generally accepted explanation of resistance. First, in susceptible reactions the lesion is usually considerably larger than the area occupied by the parasite, which is the part of the lesion formed first. Why does the parasite not occupy the whole of the killed tissue? Why also is this usually quite limited in extent? Also, most host-selective toxins, apart from their high selectivity, are also very toxic. This means that while very small amounts will indeed kill susceptible cells, the amounts that kill resistant cells are also small by the usual standards. It needs to be shown, therefore, that the amounts in infected resistant tissue are indeed too small. And, as has usually not been done, it must also be shown that there is enough to kill susceptible cells despite high toxicity. Then the responses of cells to infection must be interpreted in relation to selective toxicity.

In certain diseases one cell or a few cells are killed in the resistant response in which the parasite grows little if at all. One must now ask how these cells are killed, because their death is contrary to the hypothesis that they are killed by the host-selective toxin. And if the parasite kills a few resistant cells by other means, why does it not continue to kill more? Is it not possible that in diseases caused by parasites that produce these toxins, resistance again is induced? To pose this question is not to dispute the selectivity of the toxin or its role in the susceptible reaction, though its selectivity may now be unnecessary. But is its specificity needed to explain the resistant response? There are, in fact, many other diseases in which characteristic symptoms can reasonably be ascribed to toxins produced by parasites. These toxins are nonspecific, although the parasites that produce them are highly specific in their parasitism.

ANOTHER INTERPRETATION OF THE HYPERSENSITIVE REACTION

The hypersensitive reaction has usually been linked to resistance in the ways described above. It has also been regarded quite differently, mainly on the basis of experiments in which susceptible plant tissues are treated with substances toxic to virulent pathogens used to inoculate the tissues (13). The susceptible reaction typical of untreated tissues is replaced by cell death, similar to that which occurs in HR, and the accumulation of phytoalexins. Extracts from the parasite also killed cells and caused the accumulation of phytoalexins. Separately, neither the virulent parasite nor its inhibitor did so. From such results it is argued that the first and critical event in resistance is an interaction between parasite and plant cell that kills or damages the parasite. Substances are then released from the parasite that kill or damage the cell and cause the accumulation of phytoalexins. (In this work, and in comments on it, HR has been used, I think too loosely, for the killing of cells no matter how caused, and too much significance has been attached to killing as such.)

It may seem odd that resistant responses are so commonly accompanied by the accumulation to high concentrations of substances that are quite toxic to the parasite. But the proponents of the hypothesis argue that although the phytoalexins and, by implication, other mechanisms of resistance are quite irrelevant in resistance to the primary parasite, they may function against secondary parasites. This does not seem too plausible. Another point is that in certain diseases caused by parasitic fungi, there is good visual evidence that the parasite does continue to grow after cells have been killed. Also, parasitic fungi and bacteria can often be readily isolated from lesions in which their growth is inhibited in the dead cells. Although the precise role of HR, not too narrowly interpreted, and the accumulation of phytoalexins, lignification, and other potential mechanisms of resistance remain to be established, it is perhaps somewhat premature to dismiss them as irrelevant symptoms of resistance and in no way connected with causes (23).

NONHOST RESISTANCE

So far I have dealt almost wholly with resistance in the reactions between parasitic races and cultivars that are controlled by single genes with major effects. And I have put forth the view that resistance controlled by many genes may also be based on many single-gene reactions, each with minor effects that add up to the substantial resistance, although it is usually less than single-gene-controlled resistance. There remains another type of resistance that is probably the most important of all but about which practically nothing is known. Sometimes called nonhost resistance, the phenomenon is the resistance of plants to challenge by parasitic bacteria and fungi that do, however, cause disease of other (host) plants. This is, of course, the most common form of "resistance." Parasitism almost always is a highly specific phenomenon: a particular plant can be parasitized by comparatively few of the many thousands of bacteria and fungi that are plant parasites; and similarly, a particular parasite almost always is able to parasitize but a few of the many thousands of plants available to it. The response of a "nonhost" plant to a bacterium or a fungus that is not "its" parasite is usually very similar to HR in race-cultivar interactions, namely, the death of one cell or a few cells and very limited growth of the parasite. Or there may be no observable response, although one should not assume that there has been no response. In a few cases the response is accompanied by the accumulation of phytoalexins. Indeed, some of the earliest work on phytoalexins was based on nonhost resistance.

It is very difficult to explain nonhost resistance in terms of matching genes for avirulence and resistance because each plant is a host for only a few parasites and because each parasite is avirulent to so many plants. An alternative explanation has, in part, already been referred to. All plants respond more or less nonspecifically to one or more of a small range of substances common to parasitic bacteria and fungi; for example, the high-molecular-weight glucans or glycoproteins of fungi referred to earlier. This nonspecific response then causes the later reactions expressed as resistance. A parasite of a particular plant may therefore be parasitic precisely because it does not cause the initial nonspecific response. This, however, seems very unlikely because, of course, the parasite remains avirulent with respect to other plants. Alternatively, although it may still cause the nonspecific response, other products it produces may suppress or nullify the later reactions that would lead to resistance. Such products would act on something in the plant that distinguishes it from other plants (37). This specific suppression of the nonspecifically induced resistance mechanism establishes what has been called the basic compatibility between plant and parasite. At present, apart from the substances already referred to, we have no information on the nature of the substances that cause these effects. With parasitism established, it will now be advantageous to the plant to change in a way that would allow the parasite to induce resistance. This will in turn give a selective advantage to a change in the parasite that would make it virulent again. Within populations of plants and parasites there evolves a succession of genes for resistance matched by genes for avirulence, and molecular models for such a system have been proposed (19). There is, however, very little experimental evidence for these or other similar models. The pressing need now is for far more research on nonhost resistance to complement that being done on race-specific resistance, which now seems likely to make important advances in the near future.

REFERENCES

1. Albersheim, P., and Anderson-Prouty, A. J. (1975): Carbohydrates, proteins, cell surfaces, and the biochemistry of pathogenesis. *Annu. Rev. Plant Physiol.*, 26:31–52.
2. Anderson, A. J. (1980): Differences in the biochemical composition and elicitor activity of extracellular components produced by three races of a fungal plant pathogen, *Colletotrichum lindemuthianum. Can. J. Microbiol.*, 26:1473–1479.
3. Asada, Y., Ohguchi, T., and Matsumoto, I. (1979): Induction of lignification in response to fungal infection. In: *Recognition and Specificity in Plant Host-Parasite Interactions*, edited by J. M. Daly and I. Uritani, pp. 99–112. University Park Press, Baltimore.
4. Ayers, A. R., Ebel, J., Valent, B., and Albersheim, P. (1976): Host-pathogen interactions. X. Fractionation and biological activity of an elicitor isolated from the mycelial walls of *Phytophthora megasperma* var. *sojae. Plant Physiol.*, 57:760–765.
5. Bailey, J. A. (1981): Physiological and biochemical events associated with the expression of resistance to disease. In: *Active Defence Mechanisms in Plants*, edited by R. K. S. Wood. Plenum Press, New York.
6. Bailey, J. A., and Deverall, B. J. (1971): Formation and activity of phaseollin in the interaction between bean hypocotyls *(Phaseolus vulgaris)* and physiological races of *Colletotrichum lindemuthianum. Physiol. Plant Pathol.*, 1:435–449.
7. Bruegger, B. B., and Keen, N. T. (1979): Specific elicitors of glyceollin accumulation in the *Pseudomonas glycinea*—soybean host-parasite system. *Physiol. Plant Pathol.*, 15:43–51.
8. Callow, J. A. (1977): Recognition, resistance and role of lectins in host-parasite interactions. *Adv. Bot. Res.*, 4:1–49.
9. Daly, J. M. (1976): Specific interactions involving hormonal and other changes. In: *Specificity in Plant Diseases*, edited by R. K. S. Wood and A. Graniti, pp. 151–165. Plenum Press, New York.
10. Day, P. R. (1974): *Genetics of Host-Parasite Interactions*. W. H. Freeman, San Francisco.
11. Doke, N., Garas, N. A., and Kuć, J. (1979): Partial characterization and aspects of the mode of action of hypersensitivity-inhibiting factor (HIF) isolated from *Phytophthora infestans. Physiol. Plant Pathol.*, 15:127–140.
12. Ellingboe, A. H. (1981): Genetical aspects of active defence. In: *Active Defence Mechanisms in Plants*, edited by R. K. S. Wood. Plenum Press, New York *(in press)*.
13. Érsek, T., Barna, B., and Király, Z. (1973): Hypersensitivity and the resistance of potato tuber tissues to *Phytophthora infestans. Acta Phytopathol. Acad. Sci. Hung.*, 8:3–12.
14. Flor, H. H. (1971): Current status of the gene-for-gene concept. *Annu. Rev. Phytopathol.*, 9:275–296.
15. Garas, N. A., Doke, N., and Kuć, J. (1979): Suppression of the hypersensitive reaction in potato tubers by mycelial components from *Phytophthora infestans. Physiol. Plant Pathol.*, 15:117–126.
16. Hargreaves, J. A., and Bailey, J. A. (1978): Phytoalexin production by hypocotyls of *Phaseolus vulgaris* in response to constitutive metabolites released by damaged cells. *Physiol. Plant Pathol.*, 13:89–100.
17. Heath, M. C. (1979): Partial characterization of the electron-opaque deposits formed in the non-host plant, French bean, after cowpea rust infection. *Physiol. Plant Pathol.*, 15:141–148.
18. Heath, M. C. (1980): Effects of infection by compatible species or injection of tissue extracts on the susceptibilty of non-host plants to rust fungi. *Phytopathology*, 70:356–360.
19. Heath, M. C. (1981): The absence of active defence mechanisms in compatible host-pathogen combinations. In: *Active Defence Mechanisms in Plants*, edited by R. K. S. Wood. Plenum Press, New York *(in press)*.
20. Keen, N. T. (1981*a*): Specific recognition in gene-for-gene host parasite systems. In: *Advances in Plant Pathology*, edited by D. Ingram and P. H. Williams. Academic Press, New York *(in press)*.
21. Keen, N. T. (1981*b*): Mechanisms conferring specific recognition in gene-for-gene plant parasite systems. In: *Active Defence Mechanisms in Plants*, edited by R. K. S. Wood. Plenum Press, New York *(in press)*.
22. Keen, N. T., and Legrand, M. (1980): Surface glycoproteins: evidence that they may function as the race specific phytoalexin elicitors of *Phytophthora megasperma* f. sp. *glycinea. Physiol. Plant Pathol.*, 17:175–192.
23. Király, Z. (1980): Defences triggered by the invader; hypersensitivity. In: *Plant Disease. An Advanced Treatise. Vol. 5: How Plants Defend Themselves*, edited by J. G. Horsfall and E. B. Cowling, pp. 201–224. Academic Press, New York.

24. Klement, Z., and Goodman, R. N. (1967): The hypersensitive reaction to infection by bacterial plant pathogens. *Annu. Rev. Phytopathol.*, 5:17–44.
25. Martin, T. J., and Ellingboe, A. H. (1976): Differences between compatible parasite/host genotypes involving the Pm4 locus of wheat and the corresponding genes in *Erysiphe graminis* f. sp. *tritici*. *Phytopathology*, 66:1435–1438.
26. Müller, K. O. (1959): Hypersensitivity. In: *Plant Pathology: An Advanced Treatise*, edited by J. G. Horsfall and A. E. Dimond, pp. 469–519. Academic Press, New York.
27. Nass, H. A., Pedersen, W. L., Mackenzie, D. R., and Nelson, R. R. (1981): *Phytopathology (in press)*.
28. Ouchi, S., Hibino, C., Oku, H., Fujiwara, M., and Nakabayashi, H. (1979): The induction of resistance or susceptibility. In: *Recognition and Specificity in Plant Host-Parasite Interactions*, edited by J. M. Daly and I. Uritani, pp. 49–65. University Park Press, Baltimore.
29. Sequeira, L. (1978): Lectins and their role in host-pathogen specificity. *Annu. Rev. Phytopathol.*, 16:453–481.
30. Sequeira, L. (1980): Defences triggered by the invader: recognition and compatibility phenomena. In: *Plant Disease. An Advanced Treatise. Vol. V: How Plants Defend Themselves*, edited by J. G. Horsfall and E. B. Cowling, pp. 179–200. Academic Press, New York.
31. Sidhu, G. S. (1975): Gene-for-gene relationships in plant parasitic systems. *Sci. Prog. Oxford*, 62:467–485.
32. Tomiyama, K., Doke, N., Nozue, M., and Ishiguri, Y. (1979): The hypersensitive response of resistant plants. In: *Recognition and Specificity in Plant Host-Parasite Interactions*, edited by J. M. Daly and I. Uritani, pp. 69–84. University Park Press, Baltimore.
33. Van der Plank, J. E. (1963): *Plant Diseases, Epidemics and Control*. Academic Press, New York.
34. Wade, M., and Albersheim, P. (1979): Race-specific molecules that protect soybeans from *Phytophthora megasperma* var. *sojae*. *Proc. Natl. Acad. Sci. USA*, 76:4433–4437.
35. Ward, H. M. (1902): On the relations between host and parasite in the bromes and their brown rust *Puccinia dispersa* Erikss. *Ann. Bot. (London)*, 16:223–315.
36. West, C. A., Wade, M., McMillan, C., III, and Albersheim, P. (1980): Purification and properties of invertases extractable from *Phytophthora megasperma* var *sojae* mycelia. *Arch. Biochem. Biophys.*, 201:25–35.
37. Wood, R. K. S. (1976): Specificity—an assessment. In: *Specificity in Plant Diseases*, edited by R. K. S. Wood and A. Graniti, pp. 327–338. Plenum Press, New York.
38. Yoder, O. C. (1980): Toxins in pathogenesis. *Annu. Rev. Phytopathol.*, 18:103–129.
39. Yoshikawa, M., Madama, M., and Masago, H. (1981): Release of a soluble phytoalexin-elicitor from mycelial walls of *Phytophthora megasperma* var. *sojae* by soybean tissues. *Plant Physiol. (in press)*.
40. Ziegler, E., and Albersheim, P. (1977): Host-pathogen interactions. XIII. Extracellular invertases secreted by three races of a plant pathogen are glycoproteins which possess different carbohydrate structures. *Plant Physiol.*, 59:1104–1110.

Molecular Biology of Parasites, edited by
J. Guardiola, L. Luzzatto, and W. Trager.
Raven Press, New York © 1983.

Genetic Colonization of Plants by Bacteria: *Agrobacterium tumefaciens*

L. Willmitzer* and J. Schell*†

**Max-Planck-Institut für Züchtungsforschung, Köln-Vogelsang, Federal Republic of Germany; and †Laboratorium voor Genetika, Rijksuniversiteit, Ghent, Belgium*

Agrobacterium tumefaciens form a group of gram-negative soil bacteria that induce so-called crown-gall tumors on most dicotyledonous plants. Accumulating evidence has been obtained in several laboratories demonstrating that a group of large plasmids, called Ti-plasmids, are involved in this neoplastic transformation of plant cells.

Since the discovery of the Ti-plasmids, several important observations have allowed a fairly precise explanation of the mechanism underlying the transformation of plant cells. First of all, it was demonstrated that Ti-plasmids carry genes that somehow determine the specificity of synthesis of low-molecular-weight compounds (so-called opines) by the transformed plant cell. Furthermore, Ti-plasmids were shown to harbor genes that allow *Agrobacteria* to use these opines specifically as their sole carbon, nitrogen, and energy source. These observations provided genetic evidence in favor of a model involving part of the Ti-plasmid to be transferred from bacterium to plant.

Molecular hybridization techniques have recently been demonstrated that such a transfer of part of the Ti-plasmid actually occurs. The T-DNA (i.e., the Ti-plasmid DNA segment present in transformed plant cells) corresponds to one continuous segment of the Ti-plasmid with a molecular weight of 10 to 15 $\times$ 10^6. It is present in the nucleus of the transformed plant cell, where it is integrated into plant DNA.

The T-DNA is transcribed in the plant cell to give rise to seven distinct RNAs of different lengths, which originate from different regions of the T-DNA. The RNAs are polyadenylated and thus have properties typical of eukaryotic mRNAs. The transcription of the T-DNA is inhibited by low concentrations of α-amanitin, suggesting that the host RNA polymerase II is responsible for the transcription of the T-DNA.

In vitro translation of T-DNA specific mRNAs gives rise to several distinct protein bands on sodium dodecyl sulfate-polyacrylamide gels. The molecular weight of one of the proteins coincides with the molecular weight of the enzyme synthesizing octopine (a member of the opines) in the transformed plant cell. This is strong evidence that the T-DNA codes for the enzymes necessary for opine synthesis in the transformed plant cells.

Genetic techniques (insertion and deletion mutants) have established that the T-DNA consists of at least three different functional units. A central segment of molecular weight about 4.5×10^6 is conserved in all types of Ti-plasmids. Mutations in this segment invariably result in the loss of capacity to produce plant tumors. To either side of this conserved DNA segment, we found DNA sequences that are different in different types of Ti-plasmids. Mutations in these nonconserved segments of the T-DNA do not result in the loss of capacity to induce tumors. Mutations in what has been called the right side of the T-DNA ($\pm 1.5 \times 10^6$ molecular weight) do, however, result in Ti-plasmids that induce tumors in which no opine synthesis occurs, thus identifying the DNA sequence involved in opine synthesis.

The demonstration that the Ti-plasmid DNA segment specifying nopaline synthesis in transformed plant cells is part of the T-DNA but not functionally required for the maintenance or induction of the neoplastic state, supports the genetic colonization model of crown gall. Recent results have shown that the Ti-plasmid can be used as a vector for the experimental introduction and stable maintenance of foreign genes in higher plants.

REFERENCES

1. Braun, A. C. (1978): Plant tumors. *Biochim. Biophys. Acta*, 516:167.
2. Chilton, M. D., et al. (1977): Stable incorporation of plasmid DNA into higher plant cells: the molecular basis of crown gall tumorigenesis. *Cell*, 11:263–271.
3. Lemmers, M., et al. (1980): Internal organization, boundaries and integration of Ti-plasmid DNA in nopaline crown gall tumours. *J. Mol. Biol.*, 144:355–378.
4. Schell, J., et al. (1979): Crown gall bacterial plasmids as oncogenic elements. In: *Molecular Biology of Plants*, edited by I. Rubenstein, R. L. Phillips, C. E. Green, and B. G. Gengenbach. Academic Press, New York.
5. Zaenen, I., et al. (1974): Supercoiled circular DNA in crown-gall inducing *Agrobacterium* strains. *J. Mol. Biol.*, 86:109.

Molecular Biology of Parasites, edited by
J. Guardiola, L. Luzzatto, and W. Trager.
Raven Press, New York © 1983.

Parasitism Versus Symbiosis: Interaction Between *Rhizobium* and Leguminous Plants

Marco P. Nuti

Istituto di Chimica, Industrie e Microbiologia agraria, Università di Padova, 35100 Padova, Italy

The members of Rhizobiaceae are gram-negative soil microbes capable of entering unique interactions with plants, e.g., inducing formation of root and stem nodules, tumorigenic crown galls, and hairy roots. Among Rhizobiaceae, *Rhizobia* can shift from endoparasitic to endosymbiotic forms, the former being the invasive type that enter through the infection threads, the latter the so-called bacteroids entrapped in the nodules. Endoparasitic bacteroids are also known, i.e., those unable to fix atmospheric dinitrogen in the nodule; whereas the *Rhizobia* within the infection threads and *nod*$^+$*fix*$^-$ bacteroids are entirely dependent on the nitrogen supply from the legume, *nod*$^+$*fix*$^+$ bacteroids do provide the plant with substantial amounts of fixed nitrogen.

Earlier genetic studies (1,2) suggested that genes controlling or determining symbiotic functions (*sym* genes) could have an extrachromosomal location in *Rhizobium*. By using strongly polar detergents that allow dissociation of plasmid DNA from membrane complexes, it has become clear that all members of Rhizobiaceae harbor extrachromosomal elements of unusually large size, 90 to 300 $\times$ 10^6 or more (3–7), which had been overlooked in earlier biochemical investigations.

In *Rhizobia*, up to 20% or more of the total genetic information is plasmid-encoded, and the involvement of large plasmids in the symbiosis with legumes has been further substantiated by various independent biochemical and genetic evidence:

1. The isolation of symbiotically defective mutants is concomitant with the loss of a plasmid (4) or with physically detectable change such as a large deletion in one of the indigenous plasmids (8). Recently, the isolation of *nod*$^-$ strains from *nod*$^+$*fix*$^+$ *R. leguminosarum*, *nod*$^+$ *R. trifolii*, and *nod*$^+$ *R. meliloti* has been reported (4,9–11).

2. The instability of symbiotic properties in *Rhizobia* is well documented. As reviewed elsewhere (12–20), this argument led to contradictory results, especially when acridine dyes were used to induce plasmid loss in *Rhizobia*.

3. Under stringent conditions it is possible to transfer symbiotic phenotypes at frequencies greater than those found for chromosomal markers (21); for instance,

the ability to nodulate peas can be transferred at high frequency from *R. leguminosarum* to other rhizobial species that nodulate bean or clover (8). More recently, using *R. trifolii* PHR865 (pRtr5a::Tn5) as a donor, it has been possible to transfer nodulation to several *nod*$^+$ *fix*$^+$ and *nod*$^-$ *R. leguminosarum* and to restore the *nod*$^+$ *fix*$^+$ phenotype on clover in the transconjugants (13,18). Furthermore, of particular relevance is the transfer of pRtr5a::Tn5 from the same donor to Ti-plasmid-cured strains of *A. tumefaciens* that became *nod*$^+$ on clover; the transconjugants were able to attach to the clover root hairs, to induce marked curling and Shepherd's crook formation, to establish normal infection threads, and to invade the plant cortical cells; although envelopes are formed around *Agrobacterium* cells, these do not undergo transition to bacteroids. Since the same bacterium could be reisolated from the nodules and used to induce nodule formation on the same host plant, Koch's postulates have been met. The new hybrid species also shows the same plasmid pattern, i.e., the cryptic *Agrobacterium* plasmid and pRtr5a, after reisolation. These results provide strong evidence that the genetic determinants of Roa, Hac, Inf, Noi, and Bar (12,14) are plasmid-borne.

4. In *Rhizobia*, as in *Agrobacterium*, more than one plasmid is usually present in a given strain; though helpful for strain characterization within some species, overall plasmid(s) topology has not yet been considered a taxonomically relevant tool.

Earlier studies involving hybridization between Southern blots of Eco RI-digested *Rhizobium* plasmid(s) DNA (15) or total *Rhizobium* DNA (16) and small amplifiable plasmids carrying *nif* structural genes of *K. pneumoniae* showed substantial homology at specific sequence levels. To circumvent ambiguities deriving from the use of *Rhizobium* plasmid "pools," these studies have been extended by using single plasmid preparations. *Nif* structural genes homologous to *K. pneumoniae* have now been assigned to specific plasmids of *R. trifolii, R. leguminosarum*, and *R. phaseoli*, e.g., pRtr5*a*, pRle1001*a*, and pRph3622b, respectively. These genes could not be observed on plasmid isolated from *R. meliloti*, e.g., pRmeV7*a*, though they could be present on a larger plasmid as yet unidentified, due to methodological limitations. In pRle1001*a* and pRtr5*a* the structural genes for nitrogenase appear to be clustered in one Eco RI fragment, and in pRph3622 they are distributed on three Eco RI fragments. Further cross-hybridization studies clearly show that these genes are part of a DNA sequence conserved in *R. leguminosarum, R. trifolii*, and *R. phaseoli*; within the above species the homology with *nif* H is much less than with *nif* D, the latter being the most conserved gene (22).

5. Plasmid genes are actively transcribed in the nodules. By hybridization of ^{32}P-RNA from broth-cultured *Rhizobia* or endosymbiotic bacteroids and plasmid "pools" (17) or single plasmids of *R. leguminosarum*, specific fragments can be identified as transcribed in the bacteroids; these studies also indicate that not only structural genes for nitrogenase but other genes involved in symbiosis are located on indigenous *Rhizobium* plasmids.

REFERENCES

1. Higashi, S. (1976): *J. Gen. Appl. Microbiol.*, 13:391–403.
2. Dunican, L. K., O'Gara, F., and Tierney, A. B. (1976): In: *Symbiotic Nitrogen Fixation in Plants*, edited by P. S. Nutman, pp. 77–90. Cambridge University Press, Cambridge.
3. Nuti, M. P., Ledeboer, A. M., Lepidi, A. A., and Schilperoort, R. A. (1977): *J. Gen. Microbiol.*, 100:241–248.
4. Prakash, R. K., et al. (1980): In: *Nitrogen Fixation*, edited by W. E. Newton and W. H. Orme-Johnson, pp. 139–163. Madison University Park Press, Baltimore.
5. Casse, F., Boucher, C., Julliot, J. S., Michel, M., and Denarie, J. (1979): *J. Gen. Microbiol.*, 113:229–242.
6. Moore, L., Warren, G., and Stroebel, G. (1979): *Plasmid*, 2:617–626.
7. Gross, D. G., Vidaver, A. K., and Klucas, R. V. (1979): *J. Gen. Microbiol.*, 114:257–266.
8. Johnston, A. W. B., Beringer, J. E., Beynon, J. L., Brewin, N., Buchanan-Wollaston, A. W., and Hirsh, P. (1979): In: *Plasmids of Medical, Environmental, and Commercial Importance*, edited by K. N. Timmis and A. Puhler, pp. 317–325. Elsevier/North-Holland Biomedical Press, Amsterdam.
9. Lepidi, A. A., Nuti, M. P., Bagnoli, G., Filippi, C., and Galluzzi, R. (1979): In: *Some Current Research on* Vicia faba *in Western Europe*, edited by D. A. Bond, G. T. Scarascia-Mugnozza, and M. H. Poulsen, pp. 436–460. Off. Publ. Eur. Comm.
10. Zurkowski, W., and Lorkiewiez, Z. (1979): *Genet. Res.*, 32:311–315.
11. Casadesus, J., Ianez, E., and Olivares, J. (1982): *Mol. Gen. Genet. (in press).*
12. Nuti, N. P., Lepidi, A. A., Prakash, R. K., Hooykaas, P. J. J., and Schilperoort, R. A. (1980): In: *Molecular Biology of Plant Tumors*, edited by G. Kahl and J. Schell, pp. 561–587. Academic Press, New York.
13. Hooykaas, P. J. J. (1980): Presented at the EEC Meeting, Norwich, September.
14. Vincent, J. M. (1980): In: *Nitrogen Fixation*, edited by W. E. Newton and W. H. Orme-Johnson. Madison University Park Press, Baltimore.
15. Nuti, M. P., Lepidi, A. A., Prakash, R. K., Schilperoort, R. A., and Cannon, F. C. (1979): *Nature (London)*, 282:533–535.
16. Ruvkun, G. B., and Ausubel, F. M. (1980): *Proc. Natl. Acad. Sci. USA*, 77:191–195.
17. Krol, A. J. M., Hontelez, J. G. J., Van Den Bos, R. ., and Van Kammen, A. (1980): *Nucleic Acids Res.*, 8:4337–4367.
18. Hooykaas, P. J. J., Van Brussel, A. A. N., Den-Dulk-Raas, H., Vanslogteren, G. M. S., and Schilperoort, R. A. (1981): *Nature*, 291:351–353.
19. Prakash, R. K. (1981): Ph.D. Thesis, Leiden.
20. Denarie', J., Boistard, P., Casse-Delbert, F., Atherly, A. G., Berry, J. O., and Russel, P. (1981): In: *The Biology of Rhizobiaceae*, edited by K. L. Giles and A. G. Atherly. Academic Press, New York.
21. Johnston, A. (1981): In: *Recent Developments in Nitrogen Fixation*, edtied by W. Newton and A. H. Gibson, pp. 163–164. Australian Academy of Sciences Publications, Canberra.
22. Prakash, R. K., Schilperoort, R. A., and Nuti, M. P. (1981): *J. Bacteriol.*, 145:1129–1136.

Molecular Biology of Parasites, edited by
J. Guardiola, L. Luzzatto, and W. Trager.
Raven Press, New York © 1983.

DNA Structure and Characterization

C. Frontali and E. Dore

Laboratorio di Biologia Cellulare e Immunologia, Istituto Superiore di Sanità, Rome 00161, Italy

STRUCTURE AND TOPOLOGY

The title of a recent paper by Crick et al. (7) is striking: "Is DNA really a double helix?" Since everyone is familiar with the double-helix model, the question mark in this title is somewhat disturbing. The brief note by Watson and Crick (26) that appeared in *Nature* in 1953 is something of a landmark in molecular biology. Yet, as Crick (6) himself recognized, a critical reading of the initial communication hardly supports its status—it contains very little substantiating experimental evidence. The model proposal was based not so much on the x-ray diffraction data collected by the King's College group (29), as on steric, biochemical, and genetic considerations. The great achievement by Watson and Crick was certainly due to their capacity to integrate information of quite different nature, such as the equimolarity of adenine and thymine and of guanine and cytosine, the genetic role recognized for DNA, and the necessity of postulating a self-replicating structure. It has to be emphasized that the proposed model gave a sufficiently good agreement between calculated and experimental diffraction patterns (28) but certainly was not the only model yielding good agreement.

This is so because fiber diffraction patterns do not allow high resolution and do not uniquely define the tridimensional structure. Good agreement could also be found, for instance, for a left-handed helical arrangement in the case of the B-form DNA (which prevails at high relative humidity). The same was not true for the A-form (observed at low relative humidity): the ready interconvertibility of the two forms was the only argument in favor of the right-handed helix proposed by Watson and Crick.

Watson and Crick's attitude was also very fruitful in that they were not put off by some very puzzling aspects of their model: the proposal of plectonemic (and not paranemic) helices in effect presents topological difficulties in replication, which we shall briefly discuss later in this chapter.

During the 28 years since Watson and Crick's original proposal, periodic attacks on their model alternated with brilliant proofs of its validity, such as the demonstration of semiconservative replication (19) and the dissociation of the two strands by treatments (heat, acid or alkaline pH) to break the hydrogen bonds *(denaturation)*, followed by reforming these bonds and thus recovering the native, double-stranded structure (*renaturation* or *reassociation*—a process we shall illustrate in detail later in this chapter.)

Antiparallelism of the two strands was first confirmed by direct evidence on the frequency of contiguous base pairs and subsequently by exact sequence determinations, ever increasing in number. Also, the linear density (mass per unit length) of DNA in solution is in reasonable agreement with Watson and Crick's B-form DNA. Now, almost 30 years after the original proposal, Watson and Crick's model still has the honor of receiving polemic attacks: Sasisekharan et al. (21,22) in 1978 suggested a different model, in which paranemic strands are helically wound, with alternating right- and left-handed turns, and claim that the agreement with diffraction data is at least as good, if not better, than for the Watson and Crick model. However, the later model proved to be unrealistic (7) on the basis of determinations of the linking number, α, possible in the case of circular DNAs [α is defined as the (integer) number of turns by which each strand is wound around the complementary one].

Circular DNA molecules exhibit the possibility of supercoiling. This superhelical winding takes place when the linking number, α, is smaller than in normal B configuration (1 turn/10 residues) and is maintained at this lower value (typically 475 turns instead of 500 in a circle 5000 base pair long) by the closure condition. Only when the closure condition is removed, e.g., by nicking one strand, does circular DNA recover the B-form, with $\alpha = 500$ (relaxed form). But DNA in solution has a conformational energy minimum corresponding to the B-form: so circular DNA deficient in primary helical turns will tend to twist further, to minimize conformational energy. This can be achieved without changing α (which in closed DNAs is a topological invariant) if negative superhelical turns are simultaneously introduced. This may be expressed by the relation $\alpha = \beta - \tau$ or linking number (invariant in closed DNAs) = actual twist number minus the number of superhelical turns (or writhing number); $475 = 500 - 25$. In relaxed circular forms, the primary helix deficit is removed, $\tau = 0$ and $\alpha = \beta$.

It is important to realize that β and τ are numbers (not necessarily integers) that are complementary in the sense just discussed: in a covalently closed, circular structure, the one cannot change without a concomitant change in the other. If for some reason β decreases, τ must decrease by the same amount, while α must be conserved. That is, turns can be shifted by this mechanism from a primary helix to a superhelix, as in a twisted helical rope.

There are two important considerations to be aware of in this connection: the first concerns replication of circular DNA; the second, preferential interaction with drugs.

In order to separate the two daughter molecules during replication of a closed, double-stranded circle, the action of a topoisomerase is needed, in order to decrease, turn by turn, the linking number. But α cannot be reduced to zero; otherwise the two strands would come apart before the DNA polymerase had accomplished its job. So topoisomerase must act until the two complete daughter molecules are linked once. At this moment, the action of a nicking enzyme is needed in order to separate the daughter molecules. If this step did not occur, the two daughter molecules would remain linked in the same fashion as with a single-turn twisted and sealed strip of paper. This is a possible explanation for the families of minicircles in kinetoplast DNA of trypanosomes.

Another peculiarity of circular DNA concerns its preferential interaction with planar molecules, like ethidium bromide, which are able to intercalate between adjacent base pairs, increasing their distance, and reducing the primary winding (expressed by the twisting number, β). In a circular DNA this process is favored by the possibility of simultaneously reducing the writhing of the helical axis, τ. So by increasing the dye:nucleotide ratio, one first observes the disappearance of superhelical turns, until a relaxed, rather rigid conformation is reached. A further increase introduces reverse superhelical turns. The sedimentation behavior typically reflects this kind of interaction: sedimentation coefficient first decreases (being more compact, the supercoiled circle has a faster rate than does a relaxed molecule) and then increases again to the original value when dye:nucleotide ratio is increased. A similar behavior is observed in the case of circular DNA interacting with the antischistosomal drug hycanthone, as well as with cloroquine. These drugs are therefore supposed to act by intercalating, although the evidence is inconclusive (25). The typical inversion of sedimentation coefficient is not present in the case of the trypanocide diminazene aceturate (Berenil®), although this drug also shows preferential interaction with circular kinetoplast DNA and with the replicating form of phage ϕX174 DNA. It is hardly necessary to emphasize that intercalating agents often induce errors in replication and are likely candidates (when not demonstrated agents, as in the case of hycanthone) for mutagenesis and possibly carcinogenesis.

Finally, it should be mentioned that it has recently been possible to obtain single crystals of double-stranded oligonucleotides with the known sequence CGCGCG and to perform a complete structural study (24) by x-ray diffraction at 1 Å, i.e., at a resolution much greater than allowed by fiber diagrams. Surprisingly, the now unambiguously determined structure differs from Watson and Crick's model in many important aspects. It consists, as proposed by Watson and Crick, of two plectonemic strands, but these are wound around each other following a left-handed, not a right-handed, helix. Furthermore, the backbone of each strand exhibits a characteristic zigzag path, and the position of bases, still contained in planes almost perpendicular to the helix axis, is different than in Watson and Crick's model. It has still to be demonstrated whether the new structure (Z-form) is stable in longer DNA stretches and in an aqueous environment, and whether its occurrence is sequence specific. The possibility of short stretches assuming different conformations would be important in providing regulatory signals.

PHYSICOCHEMICAL CHARACTERIZATION OF DNA

A number of physicochemical parameters—such as buoyant density, ultraviolet (uV) extinction coefficient, and melting temperature—and geometric-molecular parameters—such as molecular weight, radius of gyration, persistence length—have been used to characterize nucleic acid molecules; their enumeration would be rather dull, without a description of the techniques involved, but the space available here is certainly insufficient to give even a short account of the techniques more commonly employed. The development of hydrodynamic techniques (sedimentation velocity and sedimentation equilibrium in density gradients, viscosity, flow birefringence) and optical techniques (light scattering, circular dichroism, uV spectrophotometry) fostered important achievements in DNA characterization in the decade from 1955 to 1965. In the late 1960s, electron-microscopic techniques were developed, allowing direct visualization of DNA, and these contributed importantly to our knowledge concerning more exact molecular weights, circularity, and supercoiling.

In elucidating this latter property, electrophoretic techniques on agarose gels (14) were extremely important in demonstrating the quantitized behavior typical of a property (the linking number α) that must assume integer values. This technique proved to be an extremely valuable tool in studying subtle changes in DNA geometry: it has been possible to determine in this way the precise value of the average helix rotation angle per base pair (2,10) and to give an elegant demonstration that the linking number of covalently closed PM2 DNA molecules was in agreement with Watson and Crick's plectonemic model (7,16). It is interesting to note how very fine structural details can be revealed by this relatively unsophisticated technique. In the following we are primarily interested in physicochemical methods characterizing DNA for a different reason: that of discovering the amount of genetic information contained in a genome. In this regard, we have to discuss in some detail the physical chemistry of the denaturation (melting) and renaturation of DNA.

Melting

If one gradually increases the temperature of a DNA solution and monitors optical changes in the near-uV, nothing will happen (spectra recorded between 320 and 220 nm will be unchanged) until a critical temperature range is reached. Here, the absorbance A [defined as $A = \log (I_0/I)$, where I is the intensity of light transmitted by the DNA sample and I_0 is the intensity of incident light] rapidly increases, while the absorption spectrum is slightly shifted towards longer wavelengths. These changes take place in a rather restricted temperature range, beyond which the spectrum is again stable under further heating. At 260 nm (the absorption maximum of native DNA), the relative increase $\Delta A/A$ reaches a plateau value of about 40%, the exact value depending on base composition and on salt concentration in the solvent. The increase in absorbance, the "hyperchromic effect," is readily explained as due to melting of the hydrogen bonds connecting complementary base pairs and to the consequent increased exposure of the chromophores (the aromatic rings of the bases) to incident light. The bases are in effect partially shielded in native DNA, which is *hypochromic* with respect to a nucleotide mixture having the same concentration and composition. A similar cooperative absorbance increase in a narrow transition region is observed if, at constant temperature, pH is shifted to acid or alkaline values.

Once completed, the transition becomes *irreversible*; on cooling to room temperature (or going back to neutral pH), absorbance values of the native DNA are not recovered. The transition is reversible if the temperature increase (or pH shift) is reversed before the hyperchromic plateau values are reached. Together with other physicochemical (hydrodynamic) evidence, this behavior is now firmly established to correspond to melting of the bonds stabilizing the double-helical structure and to an unwinding of the two strands, which (with the exception of circularly closed DNA) are recovered in the single-stranded state when the solution is rapidly brought to initial conditions after exposure to overmelting conditions. Since α cannot change in covalently closed, circular DNA, the two strands remain linked, and a rapidly sedimenting, collapsed structure is obtained. The transition is cooperative, in the sense that a larger thermal fluctuation is needed to break one hydrogen bridge in a sequence of ordered bonds than in the case of an isolated bond. In other words, a series of, let us say, 10 bridged base pairs is more stable than 10 randomly spaced, bonded base pairs. Because of the presence of two hydrogen bonds in the A-T pair, and of three bonds in the G-C pair, short AT sequences melt at lower temperatures than equally long pure GC sequences. This is reflected in a fine structure of the transition, as revealed by high-resolution, differential melting techniques. This structure is averaged out in most routine measurements. One therefore refers only to the midpoint of the transition, T_m, which will rise with decreasing A-T percentage. As a consequence of these considerations about stability, T_m values also depend on the length of duplex structure (8).

Annealing

Separated single strands can be reannealed by incubation at a temperature value (typically 25°C below T_m) at which a dynamic equilibrium is established between formation and breakage of hydrogen bonds: incubation under these conditions eventually leads to *renaturation*, longer sequences being favored by their higher thermal stability. This is again due to reciprocal reinforcement of ordered hydrogen bonds (cooperativity), larger thermal fluctuations being needed to break longer ordered structures. The physicochemical parameters (absorbance, buoyant density, and so on) characteristic of native DNA are largely recovered; however, the final product, at infinite incubation time, will be identical to native DNA only under the stringent condition of perfect integer strands. The presence of even a smaller number of nicks gives rise after denaturation to an inhomogeneous population of single-stranded fragments, and the annealing of strands of different length leads to the formation of duplexes ending with single-stranded tails. After prolonged incubation, dendritic molecules are recovered, and as renaturation proceeds, it becomes increasingly difficult to saturate the single-stranded portions. This effect limits the percentage of renaturation actually obtainable, in a way we shall discuss quantitatively a little later. Several cycles of denaturation and renaturation can be performed; the limits are only the result of the introduction of nicks.

HYBRID MOLECULES

Mixtures of single-stranded nucleic acids (DNA or RNA) of different origin may be subjected to renaturation experiments in order to reveal the presence of identical sequences. As discussed above, hybrid duplexes will exhibit a higher thermal stability the longer the homology region. In this kind of experiment, one of the components, be it RNA or DNA, is radioactively labeled. Renaturation (or, better, hybridization) is often carried out in the presence of a solid matrix, to which the cold component has been attached by some procedure. It is thus possible, after exposure to annealing conditions, to eliminate the unreacted portion of the labeled component and to evaluate the extent reassociation from the measure of radioactivity retained by the solid matrix (agarose gels, filters, etc.) after several washings. Alternatively, renaturation is carried out in the liquid phase—samples of renaturation mixtures, cooled in order to stop reassociation, are then filtered through columns (e.g., hydroxy-apatite) or filters, separating unreacted from reassociated strands.

Further information can be obtained by heating the reassociation product retained on filters or columns progressively, until duplex regions melt again and can be eluted: the T_m after this process will be the higher the longer the homology region.

Due to space limitations, I will only very briefly discuss two important applications of these techniques.

Strain Classification Based on DNA-DNA Homology

It is a well-known fact that DNA extracted from taxonomically related strains exhibits similar base composition, and therefore similar T_m values. The similarity of T_m values, on the other hand, is certainly not a sufficient criterion for relatedness, since this parameter reflects only the average base content. The technique just described allows a much subtler test of informational relatedness. The method is as follows: cold DNA from a type strain denatured and immobilized on Millipore filters is annealed in parallel experiments with homologous and heterologous labeled DNA; the extent of homology is estimated by measuring the percentage of radioactivity retained on the filter in the heterologous mixture with respect to that retained in the homologous one. The procedure is sufficiently rapid compared with the time required to carry out a complete series of phenotypic tests, and at least in the case of microorganisms, it allows for the identification of a number of sufficiently well separated *homology groups*.

It is important to stress that taxonomical groups obtained in this way do not necessarily coincide with classical ones (12). In classical taxonomy, taxa are based on similarities in phenotypically expressed properties (e.g., presence of isoenzymes or antigens). These will generally reflect a rather restricted portion of the whole genome. In taxonomy based on DNA-DNA hybridization, on the other hand, the tests reveal homology of sequences irrespective of their actual expression. It must therefore be borne in mind that with DNA-DNA homology, a completely new kind of taxonomical study was begun in the 1960s. For a rather long time this kind of classification was considered to be superior to phenotypic taxonomy, which requires a large number of biochemical or serological tests on mutually independent properties.

This claim should be tempered, in the light of more recent data. Thus, taxonomy based on the DNA-DNA homology considers DNA as a static registration of all the potentially expressible functions, actual expression being regulated by some repressor-operator mechanism. The more recent discoveries of intervening sequences and of the possibility of genetic amplification, loss, or translocation [e.g., in ciliated protozoa (1,15)] throws into question the claimed general validity of the DNA-DNA homology approach. A much more promising approach to sound taxonomy is the one involving determination of the number of amino acid substitutions in nonessential regions of evolutionally conserved proteins (3,9,17).

Identification of Particular Sequences in Complex Structures.

The availability of radioactive probes, such as purified messenger RNA, or its complementary (c) DNA, or amplified copies of cloned DNA fragments of known function, has disclosed an infinite series of applications of hybridization techniques. In order to identify chromosomal or extranuclear localization of a cloned and labeled restriction fragment, *in situ* hybridization can be monitored by autoradiography. Southern (23) and "northern" blots are examples of techniques by which specific information is localized in a complex restriction pattern. Selection of recombinant clones containing a specific portion of the genome may be done by hybridizing a suitable probe directly onto the recombinant colonies. Even a brief account of the many important achievements made possible in recent years by such methods would be too long to give here.

RENATURATION KINETICS AND INFORMATION CONTENT

Let us now go deeper into the analysis of the renaturation process, in order to better understand the molecular basis of the method that allows estimating the total informational content present in the whole genome of a given organism from the analysis of the renaturation kinetics of the total DNA. Paradoxically, such information is obtained by using deliberately fragmented DNA. Fragmentation down to 300 to 500 base pairs is generally achieved by sonication.

In order to introduce relevant quantities, let us start with the case of a population of homogeneous DNA molecules, and let us suppose them all cut *at definite sites*. Let c_0 be the total DNA concentration expressed in moles of phosphorous per liter; c_+ and c_- are the concentrations of plus and minus strands after melting (expressed in the same units). Starting from the moment at which melting conditions are removed, and separated strands are allowed to reanneal at a suitable temperature, hydrogen bonds will begin to form between couples of strands that are near each other. However, such bonds will resist thermal fluctuations only if, adjacent to the first nucleation, homology between nucleated sequences allows the formation of an ordered array of bonded base pairs. If not reinforced by contiguous bonds, the first, isolated bond will eventually break under thermal agitation. Once a successful nucleation event has taken place, subsequent zipping of homologous regions, nucleated in register, is a very rapid process: in effect it is the first step and depends on the diffusion times of the molecules in the bulk solution, which limits renaturation rate.

This is expressed mathematically by stating that the decrease $(-dc_+, -dc_-)$ in the concentration of plus and minus strands over an infinitesimal time interval, dt, is proportional to the product of actual concentrations:

$$-dc_+ = kc_+ c_- \, dt, \quad -dc_- = kc_+ c_- \, dt.$$

But at any time, $c_+ = c_- = c/2$, so that we may write

$$-dc_+ = kc_+ c_- \, dt, \quad -dc_- = kc_+ c_- \, dt.$$

Summing up these differential contributions between time zero and time t (i.e., integrating between 0 and t), we obtain

$$-\int \frac{dc}{c^2} = \int d\left(\frac{1}{c}\right) = k_2 \int dt$$

or

$$\frac{1}{c_t} - \frac{1}{c_0} = k_2 t,$$

since $1/c$ is the function whose derivative is $-1/c^2$. The last equation can be rearranged as

$$c_0/c_t = 1/s = 1 + k_2 c_0 t, \tag{1}$$

where $s = c_t/c_0$ indicates the single-stranded, nonnucleated fraction present at time t.

It is easily recognized that these equations express typical second-order kinetics: $1/s$ increases linearly with time and, at fixed time, is proportional to c_0. Experiments performed at different c_0 values can easily be compared in a plot of $1/s$ as a function of the product $c_0 t$, where the second-order constant k_2 represents the slope of the straight line (Fig. 1a). The intercept with the ordinate axis is 1.0, indicating fully denatured starting material ($c_t = c_0$ at time zero).

Equation 1 may be reversed to give (5):

$$s = \frac{c_t}{c_0} = \frac{1}{1 + k_2 c_0 t} = \frac{1}{1 + c_0 t/(c_0 t)_{1/2}}. \tag{2}$$

In the last expression, the reassociation constant k_2 has been written as $k_2 = 1/(c_0 t)_{1/2}$. Let us explain the reason for this substitution. It is usual to plot the single-stranded fraction, s, or its complement, $(1 - s)$, the renatured fraction, as indicated

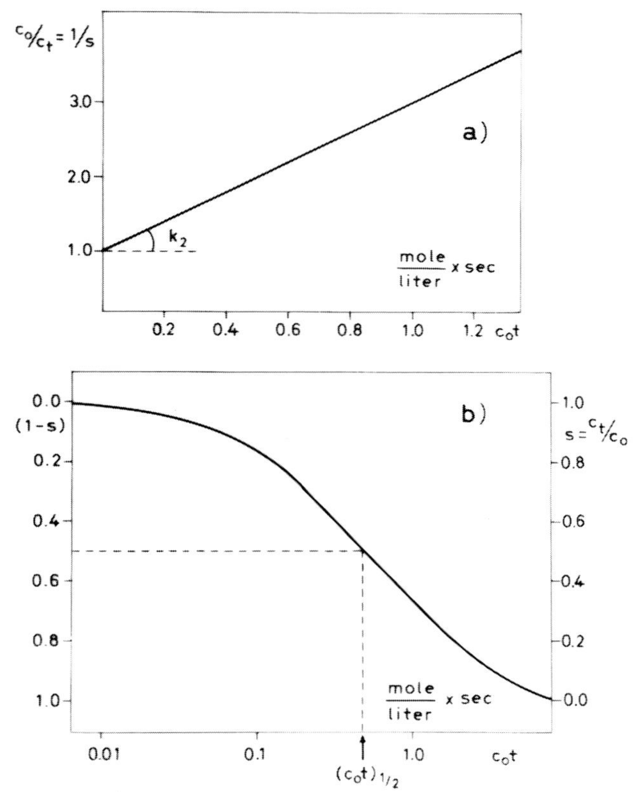

FIG. 1. A: Reciprocal second-order plot for bimolecular renaturation kinetics, representing Eq. 1. k_2 is the slope of the straight line. Intercept with ordinate axis, representing initial situation, corresponds to $c = c_0$, or $s = 1$. **B:** Fractional concentration of unreacted strands, s (see Eq. 2), and of nucleated strands $(1 - s)$ as a function of the product c_0t. The abscissa of the point corresponding to 50% renaturation is indicated as $(c_0t)_{1/2}$.

in Fig. 1b, on semilog paper. It is easy to see that $(c_0t)_{1/2}$ is the abscissa of the point where $s = ½$, i.e., where 50% of the strands have undergone renaturation.

The reassociation constant k_2, and obviously its reciprocal $(c_0t)_{1/2}$, depends on the length of the renaturing single-stranded fragments. In particular, k_2 has been shown to increase linearly with $L^{1/2}$, L being the average length of the fragments (28). Its value is also affected by the salt concentration present during incubation, increasing ionic strengths giving higher renaturation rates (28).

Now let us imagine the same renaturation experiment is carried out on two different DNA preparations, one isolated from a phage and the other from a bacterial strain. Let us suppose the two preparations have been fragmented to the same length and are incubated at the same c_0 value, all other conditions (ionic strength, temperature, etc.) being equal: it is quite evident that successful nucleation events will be less frequent [and therefore k_2 will be smaller and (c_0t) higher] for the DNA of bacterial origin, since this fragment population varies more in sequence as a result of the fragmentation procedure than does the one for phage DNA.

This fact can be generalized as follows (5,27): the reassociation constant k_2 is inversely proportional to *genetic complexity* [whereas, of course, $(c_0t)_{1/2}$ values are directly proportional to genetic complexity]. Genetic complexity is defined as the total number of base pairs present in the nonrepeating sequences building up the unique portion of the genome. It is a measure of the informational content of a genome and is expressed either as the number of base pairs (or of kilobase pairs) or molecular weight. It coincides with the number of base pairs present in the intact haploid genome only in the absence of repetitive sequences. If a repetitive fraction is present, as in most eukaryotes, genetic complexity will obviously be smaller than the haploid genome.

It is therefore possible to estimate the informational content of a given DNA, in terms of its genetic complexity, by comparing the $(c_0t)_{1/2}$ measured for its fragmented DNA with $(c_0t)_{1/2}$ values measured under identical conditions (e.g., fragment length distribution, salt concentration, temperature difference with respect to T_m, etc.) with DNAs extracted from other organisms of known genetic complexity. The relative ease of these measurements, which do not require expensive instrumentation, and the possibility of working with DNA preparations not necessarily integer explain why this procedure has rapidly become very popular. However, its widespread use has led to several misunderstandings, which will now be examined. In particular, these concern the estimate of the repetitive fraction.

Let us first suppose that repetitive sequences in the genome under study constitute a single repetition family, of repetition length equal to or multiple of the fragment length (also assuming that the fragment population is homogeneous in length or molecular weight) and that it is organized in the tandem fashion, with n identical sequences being adjacent in the integer DNA molecule. The fragmentation process in this case will bring about, besides a population of fragments containing portions of unique sequences, a class of identical fragments, whose concentration is n times greater than that of any unique fragment. This class will renature faster than the rest of the population by virtue of its n times higher concentration. A plot similar to that in Fig. 2b will be obtained, where the ratio $(c_0t)_{1/2,\mathrm{un}}/(c_0t)_{1/2,\ \mathrm{rep}}$ indicates the repetition frequency, n, and the relative amount of repetitive DNA $(=1-s_{\mathrm{un}})$ is given by the level of the first plateau. It is important to stress that there is no reason why repetitive DNA should be physically more ready capable of renaturing: the larger number of available copies is the only factor that accelerates annealing in this class.

In the reciprocal second-order plots (RSO: plot of $1/s$ vs. c_0t) in this case, a nonlinear region will be followed by linear behavior, which extrapolates back to a $1/s$ value greater than 1.0 at $t = 0$ (Fig. 2a). The linear region corresponds to pure second-order reassociation kinetics of the unique component when renaturing alone, i.e., after repetitive fragments have completed renaturation. The reciprocal intercept value therefore yields the percentage of unique fragments in the mixture, s_{un}. Obviously, the percentage of repetitive DNA is given by $(1 - s_{\mathrm{un}})$. The linear RSO

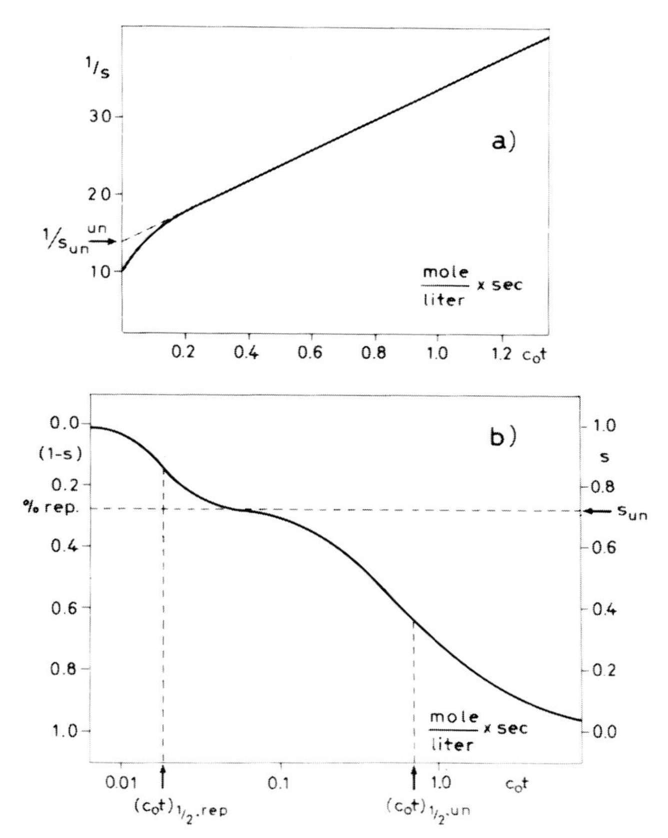

FIG. 2. A: Reciprocal second-order plot for a DNA containing repetitive sequences. Slope of the straight portion is the reassociation constant k_2 of unique DNA. Intercept with ordinate axis gives the reciprocal of s_{un}, the fractional concentration of unique fragments in the mixture. **B:** Plot of $(1 - s)$, fractional concentration of nucleated strands, as a function of c_0t. Repetitive fragments renature faster because of their higher concentration, the ratio $(c_0t)_{1/2,un}/(c_0t)_{1/2,rep}$ corresponding to the repetition frequency, n. The percentage of repetitive DNA corresponds to the intermediate plateau level.

plot often allows a more precise estimate of s_{un} than other kinds of plots. Figures 3 to 5 show examples of what we have discussed so far, in the case of *Plasmodium berghei* DNA (11).

This relatively simple kind of analysis should, in principle, give a measure not only of genome complexity, but of the percentage of repetitive DNA and of the repetition frequency, which should then allow calculation of the repetition length. Results concerning repetition frequency and length, however, and to a certain extent of the amount of repetitive DNA also, are reliable only when all these hypotheses are verified. This is seldom the case because of the following.

1. Inverted intrastrand repeats, often present, renature almost instantaneously and escape experimental observation. This amounts to saying that $s_0 < 1$, or that

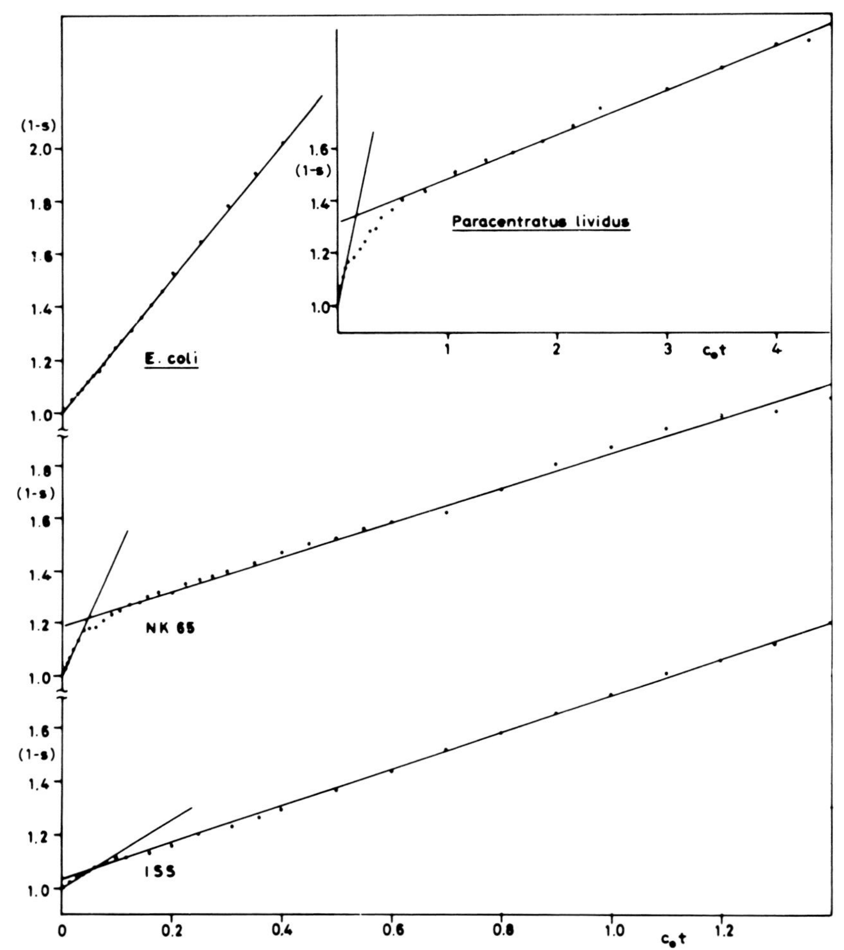

FIG. 3. Reciprocal second-order plots for DNAs from *Escherichia coli* (completely unique DNA; pure second-order kinetics), from *Paracentrotus lividus* (25% repetitive DNA), as controls, and from two strains of *Plasmodium berghei*: ISS indicates the ISTISAN strain, syringe-passaged in mice for over 30 years, that has lost the ability to give gametocytes. NK 65 strain, freshly passaged through mosquitoes, was a gift of the Wellcome Research Laboratories; it gives viable gametocytes. It may be seen that the two strains have the same reassociation constant k_2 (same slope of the linear portion) but differ in the amount of repetitive DNA. The reciprocal intercept indicates that s_{un}, the unique fraction of the genome, is 1/1.19 = 0.84 for NK 65, and 1/1.03 = 0.97 for ISS. Repetitive DNA is therefore 16% in the first case, 3% in the second.

the hyperchromic effect is incomplete. Careful measurements have to be performed at the shortest practicable times, back-extrapolation to actual zero time (removal of melting conditions) being the only means to correct for this effect.

2. DNA is generally broken by mechanical fragmentation not at definite sites, as previously supposed, but at random ones. As a consequence, homologous sequences are carried by fragments with different cuts, so that their extremities, at

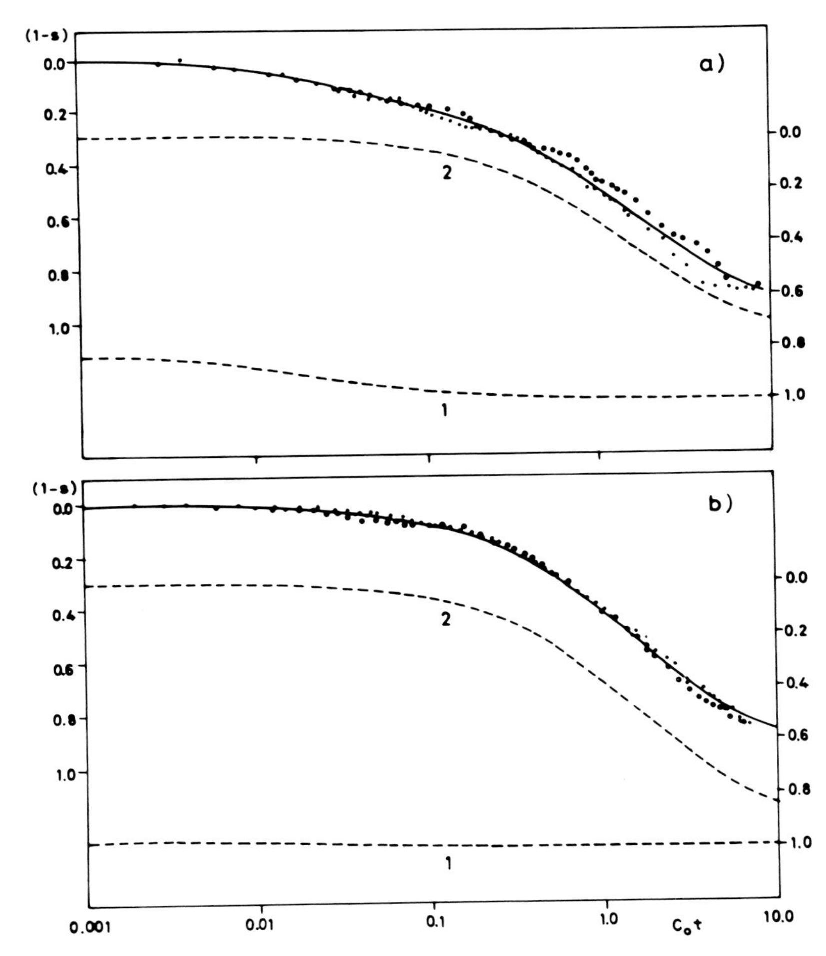

FIG. 4. Plots of (1 − *s*) vs. c_0t for *P. berghei* DNA of the NK 65 strain **(A)** and ISS strain **(B)** (see legend to Fig. 3 for strain identification). Scale to the right is again (1 − *s*) but displaced so that component curves 1 (repetitive) and 2 (unique) do not cover experimental curves. Decomposition was obtained by considering the two components as independently renaturing. Relative amounts of the two components were obtained from the analysis of RSO plots (see Fig. 3). Unique components have the same $(c_0t)_{1/2}$ value, equal to 1.52 mol s/liter.

either end of the reassociated portion, remain single-stranded. It can be calculated (27) that the average fraction, α, of base pairs found in the ordered, helical state, starting from randomly cut, single-stranded fragments of equal length L, is $\alpha = l/L = 2/3$, l being the average length of helical sections in partial duplexes.

3. Homogeneity in the length of the fragment population is clearly an oversimplification. Actual length distribution will be Gaussian, given the hypothesis of random breakage. The effect of a finite length distribution is to reduce α below the ideal value of 2/3, typically to about 0.6. A second effect of nonhomogeneous

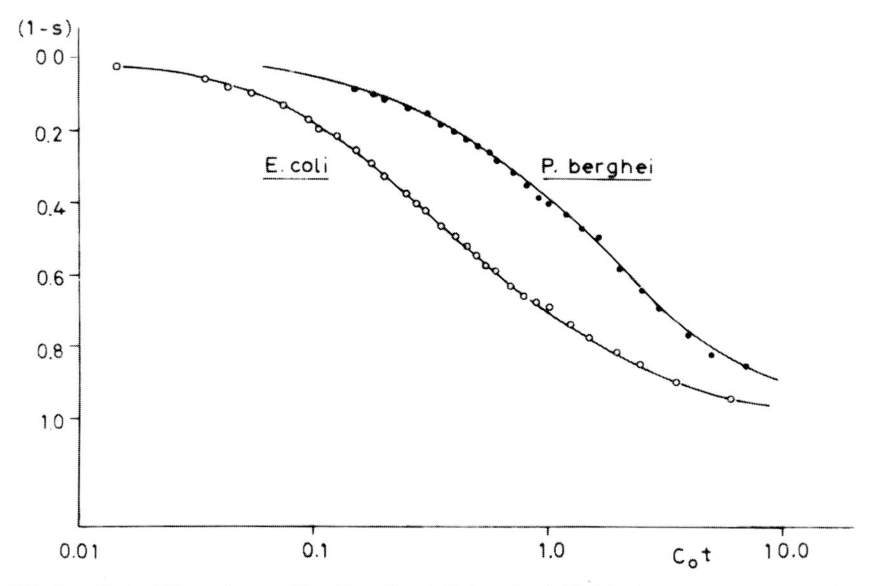

FIG. 5. Plot of $(1 - s)$ vs. c_0t for *E. coli* and *P. berghei* DNA (the latter curves report data of the unique portion of *both* strains). Comparison of $(c_0t)_{1/2}$ values (0.40 mol s/liter for *E. coli*; 1.52 mol s/liter for *P. berghei*) measured under identical conditions (1 *M* NaCl, 25°C below T_m) indicates a genetic complexity for *P. berghei* of 1.52/0.40 = 3.8 times that of *E. coli*, i.e., a genetic complexity of 1×10^{10}.

length will be that longer strands will tend to renature faster than shorter ones (since k_2 increases with $L^{1/2}$), so that RSO plots will not be linear, even for completely unique DNA, but will show a downward curvature.

4. Repetitive DNA sequences that are short with respect to L and interspersed with unique DNA do not satisfy the hypotheses thus far considered. This is the main reason why repetition frequencies and lengths are often inaccurate.

The effects mentioned in points 2 to 4 may be accounted for in more sophisticated treatments. It must be stressed that their relevance is quite different according to the actual method used to monitor renaturation kinetics. This may be done in three basically distinct ways.

First, when loaded onto hydroxyapatite columns, single-stranded DNA is eluted at low ionic strength, whereas native, renatured, or even partially renatured DNA is retained; it can be eluted at higher ionic strength. It is therefore possible to recover and measure (as a function of incubation time) fraction s, the unreacted, still single-stranded fragments present in the incubated mixture at time t. The course of renaturation can be arrested at any time t by diluting and chilling incubated samples.

Second, S1 nuclease specifically degrades single-stranded DNA and has, under optimal conditions, only a limited effect on double-stranded DNA. Its use allows measurement (as a function of incubation time) of the fraction $\alpha(1 - s)$ (the average fraction of paired nucleotides in nucleated duplexes × percent of nucleated duplexes in the mixture at time t) of nucleotides that are found in the ordered, helical state at time t.

Third, spectrophotometry, in contrast to the preceding techniques, allows continuous monitoring of reassociation kinetics. DNA fragments are melted and reannealed in a spectrophotometer cuvette, properly thermostated. Absorbance A (at 260 nm) can be recorded continuously: over time, A will decrease from the initial value $A_0 = 1.4A_{nat}$ (A_{nat} being the absorbance of native DNA, directly proportional to c_0) because of the reformation of the screening interactions typical of ordered double-stranded structure and will level off at some A_∞ value, which, as we shall see in a moment, is somewhat higher than A_{nat} because of effects 2 and 3 described above. At any time t, the actual absorbance A_t (or better, its excess, $A_t - A_\infty$) over the asymptotic level will reflect the presence of unpaired nucleotides, whether they belong to unreacted (nonnucleated) fragments or to single-stranded tails in partial duplexes. The ratio $S = (A_t - A_\infty)/(A_0 - A_\infty)$, expressing the fractional concentration of unpaired nucleotides, therefore, is not equivalent to s, the fractional concentration of unreacted fragments. The same happens in the case of S1 monitored kinetics, while chromatography on hydroxyapatite columns directly measures s. Now, the analytical treatment is different when carried out in terms of concentrations of paired and unpaired nucleotides rather than of reacted and unreacted strands. We will now develop the relevant equations, following the treatment by Rau and Klotz (20).

Let us first evaluate, at time t, the concentration S_t of unpaired nucleotides present in single-stranded fragments or tails and the concentration D_t of paired nucleotides present in complete or partial duplexes. The analogous quantities, in terms of reacted and unreacted strands, are c_t and its complement $c_0 - c_t = d_t$. If L is the length (expressed in base pairs) of the fragments, supposed to be homogeneous in molecular weight, each nucleated duplex will form, on average, αL base pairs. Thus,

$$D_t = \alpha L d_t = C_0 - C_t,$$

with $C_0 = Lc_0$ being the total nucleotide concentration.

As we already know, integration of $dc/dt = -k_2c^2$ gives

$$\frac{1}{c_t} - \frac{1}{c_0} = k_2t \quad \text{or} \quad \frac{1}{c_0 - d_t} - \frac{1}{c_0} = k_2t.$$

We shall now rewrite this equation with the new symbols:

$$\frac{1}{(C_0/L) - (D_t/\alpha L)} - \frac{1}{C_0/L} = k_2t \quad \text{or} \quad \frac{\alpha L}{\alpha C_0 - D_t} - \frac{L}{C_0} = k_2t$$

and, introducing $k'_2 = k_2/L$,

$$\frac{\alpha C_0}{\alpha C_0 - D_t} - 1 = k_2'C_0t \quad \text{or} \quad \frac{\alpha C_0}{\alpha C_0 - (C_0 - C_t)} = 1 + k_2'C_0t.$$

Finally,

$$\frac{\alpha C_0}{C_t - (1 - \alpha)C_0} = 1 + k_2' C_0 t. \tag{3}$$

Equation 3 is comparable to Eq. 1, reducing to it in the case of $\alpha = 1$, i.e., in the case of perfect reassociation.

We can rearrange Eq. 3 to express the ratio C_t/C_0 (or its inverse C_0/C_t), which is the quantity measured by spectrophotometry:

$$\frac{A_0 - A_\infty}{A_t - A_\infty} = \frac{C_0}{C_t} = 1 + \frac{\alpha k_2' C_0 t}{1 + (1 - \alpha)\, k_2' C_0 t}.$$

In other words, RSO plots are no longer linear, nor the kinetics purely second-order, unless α is equal to 1. The unpaired nucleotide fraction

$$S = \frac{C_t}{C_0} = \frac{1}{1 + \{\alpha k_2' C_0 t/[1 + (1 - \alpha) k_2' C_0 t]\}} = \frac{1 + (1 - \alpha) k_2' C_0 t}{1 + k_2' C_0 t} \tag{4}$$

also reduces to the corresponding Eg. 2, $c_t/c_0 = 1/(1 + k_2 c_0 t)$, only if $\alpha = 1$.

It is usual to plot its complement D_t,

$$D_t = 1 - S_t = 1 - \frac{A_t - A_\infty}{A_0 - A_\infty} = \frac{A_0 - A_t}{A_0 - A_\infty},$$

as a function of $C_0 t$ in semilog plots, as was done in Figs. 1b and 2b for s and $(1 - s)$. Genetic complexity can be estimated from the $(C_0 t)_{1/2}$ values so determined or from the initial slope $(=\alpha k_2/L)$ of RSO plots.

Equation 4 also tells us that C_t does not tend to zero for $t \to \infty$, as was the case for c_t (see Eq. 2). The quantity trend to zero when $t \to \infty$ can be seen from Eq. 3 to be the difference $C_t - (1 - \alpha)C_0$. This means that C_t tends to an asymptotic value $C_\infty = (1 - \alpha)C_0$. In absorbancy terms this may be stated by saying that $A_t - A_{\text{nat}}$ (proportional to the concentration of unreacted nucleotides) tends to $(1 - \alpha)(A_0 - A_{\text{nat}})$, the hyperchromic amount $\Delta A = (A_0 - A_{\text{nat}})$ being proportional to the total concentration of nucleotides, C_0. We may therefore write

$$A_\infty - A_{\text{nat}} = (1 - \alpha)(A_0 - A_{\text{nat}}),$$

which shows that A_∞ will coincide with A_{nat} only in the ideal case of perfect reassociation ($\alpha = 1$), whereas generally,

$$A_\infty = A_{\text{nat}} + (1 - \alpha)(A_0 - A_{\text{nat}}) \cong A_{\text{nat}} + (\Delta A/3) \cong 1.13 A_{\text{nat}}.$$

It is then evident that analysis of complex renaturation curves, in which one or more repetitive components are present, becomes a formidable task. In effect, the widely used relation (see ref. 4) underlying most computer available programs,

$$\frac{c}{c_0} = \Sigma_i \frac{f_i}{1 + n_i c_0 t/(c_0 t)_{1/2}}$$

(where f_i is the relative abundance and n_i the repetition frequency of the ith component), is clearly valid only under the hypothesis of independently renaturing components, as would be the case for a mixture of bacterial and phage DNA, as previously discussed. In the case of a genome containing repetitive fractions, the following requirements obtain: nonrandom fragmentation, with breakage points at the borders between unique and repetitive regions, and repetitive sequence length greater than fragment size, so that an integer number of fragments is contained in each repetitive stretch.

It is obvious that, in general, such hypotheses are not verified: they are approximately verified in the case of long repetitive sequences organized in the tandem manner, in a distinct portion of the genome. On the contrary, when repetitive sequences are highly interspersed with unique DNA, many fragments will contain both unique and repetitive parts. Mazo et al. (13,18) clearly demonstrated the inadequacy of the model of independent reassociation in this case: apparent $(c_0t)_{1/2}$ of the repetitive component increases with decreasing length of the interspersed repetitive sequence. Consequently, the usual computer analysis, when applied to spectrophotometrically monitored renaturation kinetics, underestimates the repetition frequency evaluated with the formula $n = (c_0t)_{1/2,\ \mathrm{un}}/(c_0t)_{1/2,\mathrm{rep}}$ and overestimates the length of short, highly interspersed repetitive sequences. It must be reminded that the same is not true for hydroxyapatite-monitored kinetics, where, however, the much more straightforward kind of analysis is compensated by the lower reproducibility of quantitative results.

With all these limitations, reassociation analysis remains a very powerful tool in characterizing an unknown genome. Correctly used, it yields reliable information on genetic complexity and the presence of a repetitive fraction. If, however, one wants to go further and analyze the interspersion pattern of repetitive sequences so revealed, one has to turn to different techniques. Among these, the most simple conceptually is observation under the electron microscope of partially renatured molecules, renatured at low c_0t values. In the case of *P. berghei*, this kind of analysis performed on the rapidly reassociating component, purified on hydroxyapatite, has given evidence of short (600 to 900 base pairs) repetitive sequences interspersed with long (several kilobases) stretches of unique DNA. Work is in progress on the biological role of these repetitions, which are most likely connected with gametogenesis.

REFERENCES

1. Ammermann, D., Steinbrück, G., von Berger, L., and Henning, W. (1974): *Chromosoma*, 45:401–429.
2. Anderson, P., and Bauer, W. (1978): *Biochemistry*, 17:594–601.
3. Britten, R. J., and Davidson, E. H. (1976): *Fed. Proc.*, 35:2151–2157.
4. Britten, R. J., Graham, D. E., and Neufeld, B. R. (1974): *Methods Enzymol.*, 29:363–418.
5. Britten, R. J., and Kohne, D. E. (1968): *Science*, 161:529–540.

6. Crick, F. H. C. (1974): *Nature*, 248:766–769.
7. Crick, F. H. C., Wang, J. C., and Bauer, W. R. (1979): *J. Mol. Biol.*, 129:449–461.
8. Crothers, D. M., Kallenbach, N. R., and Zimm, B. H. (1965): *J. Mol. Biol.*, 11:802.
9. Dayhoff, M. O. (1972): *Atlas of Protein Sequence and Structure*, Vol. 5. Natl. Biomed. Res. Found., Washington, D.C.
10. Depew, R. D., and Wang, J. C. (1975): *Proc. Natl. Acad. Sci. USA*, 72:4275–4279.
11. Dore, E., Birago, C., Frontali, C., and Battaglia, P. A. (1980): *Mol. Biochem. Parasitol.*, 1:199–208.
12. Frontali, C. (1979): *Ann. Ist. Sup. Sanita*, 15:415–422.
13. Goltsov, V. A., Mazo, M. A., Tarantul, V. Z., and Gasaryan, K. G. (1980): *J. Theor. Biol.*, 83:389–403.
14. Keller, W., and Wendel, I. (1974): *Cold Spring Harbor Symp. Quant. Biol.*, 39:199–208.
15. Lauth, M. R., Spear, B. B., Heumann, J., and Prescott, D. M. (1976): *Cell*, 7:67–74.
16. Liu, L. F., and Wang, J. C. (1975): *Biochem. Biophys. Acta*, 395:405–412.
17. Margoliash, E., Fitch, W. M., and Dickerson, R. E. (1979): In: *Structure, Function and Evolution in Proteins, Brookhaven Symposia in Biology,* Vol. 21, Brookhaven National Laboratory, N.Y.
18. Mazo, M. A. (1978): *Mol. Biol.*, 12:348–353.
19. Meselson, M., and Stahl, F. W. (1958): *Proc. Natl. Acad. Sci. USA*, 44:671.
20. Rau, D. C., and Klotz, L. C. (1978): *Biophys. Chem.*, 8:41–51.
21. Sasisekharan et al. (1978): *Nature*, 275:159–162.
22. Sasisekharan et al. (1978): *Proc. Natl. Acad. Sci. USA*, 75:4092–4096.
23. Southern, A. M. (1975): *J. Mol. Biol.*, 98:503.
24. Wang, A. J., Quigly, G. J., Kolpak, F. J., Crawford, J. L., van Boom, J. H., van der Masel, G., and Rich, A. (1979): *Nature*, 282:680–686.
25. Waring, M. (1970): *J. Mol. Biol.*, 54:247–279.
26. Watson, J. D., and Crick, F. H. C. (1953): *Nature*, 171:737–740.
27. Wetmur, J. G. (1967): Ph.D. Thesis, California Institute of Technology.
28. Wetmur, J. G., and Davidson, N. (1968): *J. Mol. Biol.*, 31:349–370.
29. Wilkins, M. H. F. (1956): *Cold Spring Harbor Symp. Quant. Biol.*, 21:75.

Molecular Biology of Parasites, edited by
J. Guardiola, L. Luzzatto, and W. Trager.
Raven Press, New York © 1983.

Monoclonal Antibodies and Applications to Parasitisms

Jana S. McBride

Department of Zoology, University of Edinburgh, Edinburgh EH9 3JT, Scotland

Antibodies are tools widely applicable to studies of structure, biology and function of molecules they specifically recognize, i.e., antigens, and for purification of such molecules. However, antibodies produced *in vivo* even in response to a purified antigen are usually very heterogeneous, and the composition of antisera produced by different individuals, or even by the same individual at different times, can vary. This heterogeneity and the consequent irreproducibility of antibody reagents has been a serious disadvantage, particularly for analysis of complex antigens. The introduction by Köhler and Milstein (14) of a method for production of monospecific homogeneous and reproducible antibodies against antigens of choice opened a new era in the use of antibodies as research and diagnostic probes.

Antibodies are products of specialized cells, B-lymphocytes. *Each clone* of these cells is restricted to synthesis of a *single* antibody species characterized by its specificity for a particular antigenic determinant and by its heavy- and light-chain isotypes. The specificity is determined by amino acid sequences within the variable (V) regions of the heavy and light chains, which together comprise the antigen binding site of antibody (1,36). The binding site is complementary to and combines with a region of antigen molecule that is termed the antigenic determinant (or site or epitope). Antigenic sites are small compared to the size of most antigenic molecules, involving around five to seven amino acid or sugar residues, and most natural macromolecular antigens carry several distinct determinants (2–4,11,35). On immunization, a number of different B-cell clones bearing the specificities that recognize these determinants are selected to produce the appropriate immunoglobulins, which are then secreted into serum. Thus, the more complex the antigen, the more heterogeneous is the response it is likely to induce. Additional antibody heterogeneity is due to the existence of several heavy-chain constant regions (C_H), which determine the isotype and which confer different biological properties to antibodies that may have the same specificity (1,37).

Normal antibody-producing cells have a limited life span, and it is not possible to establish clones of such cells in culture. Tumors of antibody-producing B-cells (myelomas) are capable of continuous proliferation *in vitro*, but they do not produce the wide range of antibodies that are required for a multitude of purposes. However, as Köhler and Milstein demonstrated, clones of specific immune lymphocytes can be immortalized by fusion with myeloma cells. The resulting hybrid cells (or hybrid myelomas or hybridomas) combine the growth characteristics of the tumor with the information for the production of a specific antibody coded for by the rearranged V and C genes of the immune cell. Individual hybridomas are easily isolated by cloning and can be used to produce monoclonal antibodies uncontaminated by any other. The principal advantages of monoclonal antibodies over conventional antisera are as follows: (a) each antibody is homogeneous, possessing only one specificity and one isotype; (b) monoclonal antibodies can be raised to conform to any requirement concerning specificity (including avidity) and biological activity, the only limitations being the extent of the V gene repertoire and the amount of effort a researcher is prepared to invest; and (c) once a hybridoma is established, the same antibody can be produced repeatedly, allowing for the reliable supply of reproducible reagents, including those only rarely found amongst conventional antisera.

HYBRIDOMA TECHNOLOGY

The basic requirements of the technique are within the reach of many laboratories: facilities for a simple type of mammalian tissue culture, availability of a suitable myeloma line, and liquid nitrogen storage for long-term preservation of cells are essential, and at least a limited animal housing facility is desirable.

It is beyond the scope of this paper to review the methodology in detail, and excellent work protocols are available in several published volumes (6,10, 12,16,19,24,33). In any case, the most effective way of learning the practical skills is a sojourn in a laboratory engaged in hybridoma production. Here, only the principles are outlined and a few comments that may be helpful are made.

The aim of the technology is to isolate a single clone of cells producing antibody with predetermined properties and perpetuate that clone indefinitely (Fig. 1). As said earlier, the capacity of a differentiated immune lymphocyte to make a specific antibody is immortalized by hybridization with a myeloma cell using polyethylene glycol (PEG) as the fusing agent. In any successful fusion experiment, many different hybrids are born, and the major task is to select those of interest and eliminate the irrevelent ones (usually the majority). The selection is achieved in three steps. First, myeloma cells that have not fused are removed, since they would otherwise overgrow the hybrids (HAT selection, see below). Secondly, the cultures that contain desirable antibodies are identified by a screening assay. In the third step, the selected hybrids are purified by cloning (and often recloning).

Choice of the Myeloma Line

Most hybrids so far have been derived from fusions between myelomas and immune cells from the same species, e.g., mouse-mouse, rat-rat, or human-human. Mouse-rat hybrids also have been made, but other interspecies combinations are usually unsuccessful, and it is recommended that a myeloma should originate from the same species as the immune cells.

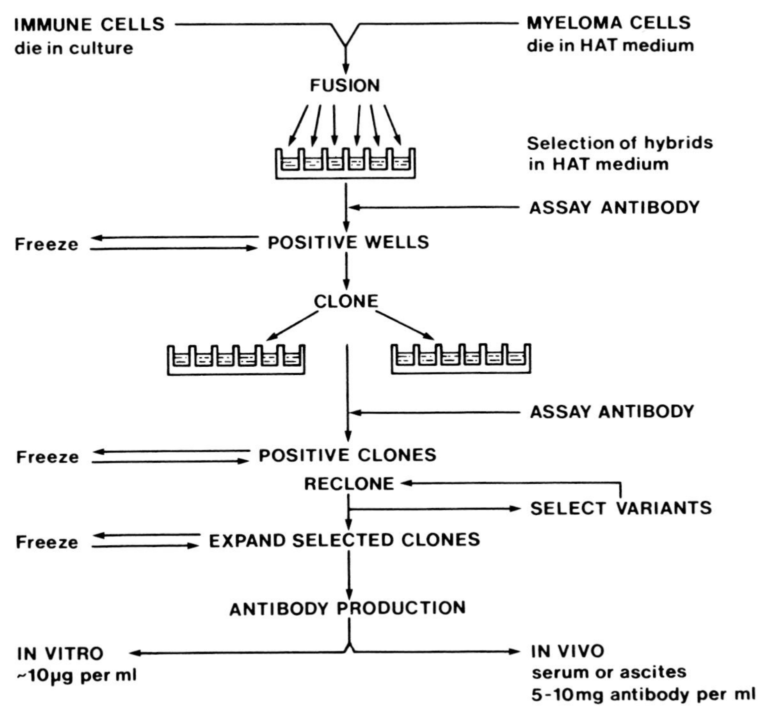

FIG. 1. Protocol for production of hybridomas (hybrid myelomas). Freeze indicates stages at which cells are frozen in liquid nitrogen for long-term storage. (Adapted from ref. 33, with permission.)

Hybrids derived from a myeloma which itself makes immunoglobulin chain(s) may secrete hybrid molecules made up of a random combination of all the chains available. Many such molecules have low avidity or are inactive (Fig. 2). Thus, it is preferable to select a myeloma line that no longer makes or secretes its own chains but that still permits secretion of the antibody determined by the normal parent.

The HAT selection (Fig. 3) against unfused myeloma cells requires myeloma mutants that lack a nucleotide salvage enzyme hypoxanthine guanine phosphoribosyl transferase (HGPRT) or thymidine kinase (TK). The enzyme-deficient mutants cannot grow in the presence of aminopterin, an inhibitor of *de novo* nucleotide synthesis, i.e., under conditions where incorporation of hypoxanthine and thymidine from the medium is essential. Hybrid cells provided with HGPRT (or TK) by the normal parental cells will survive.

To develop myeloma lines that fulfill all the above requirements, i.e., that do not produce their own immunoglobulin chains and yet confer good growth and high production and secretion phenotypes to hybrids, and that are also HAT-sensitive, is not an easy task. Mouse (13–15,32), rat (8), and human (5,23) lines so far selected have some, but not always all, of these properties. Recently, a procedure

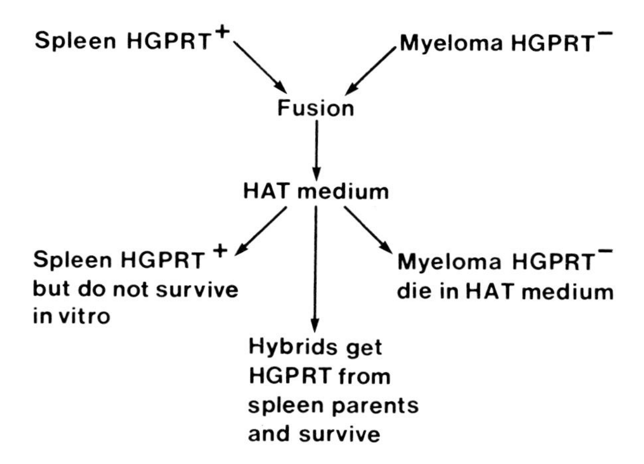

FIG. 2. The principle of HAT selection. HAT, hypoxanthine aminopterin thymidine; HGPRT, hypoxanthine guanine phosphoribosyl transferase.

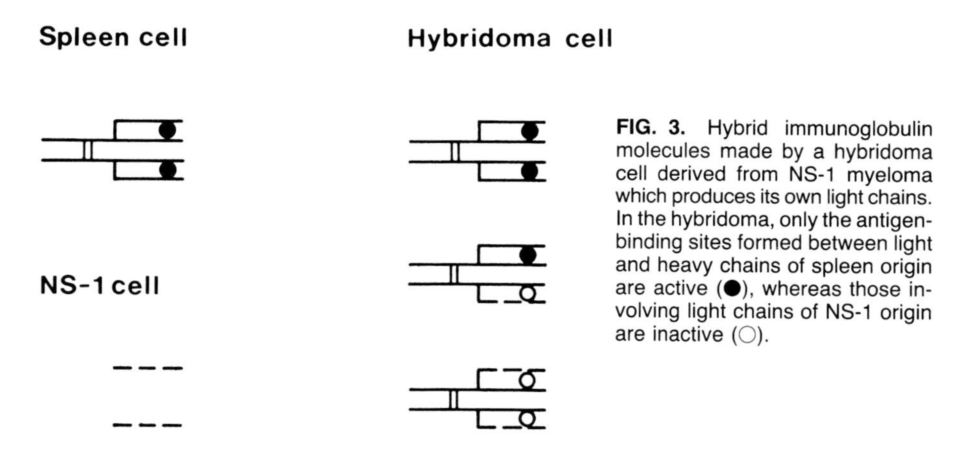

FIG. 3. Hybrid immunoglobulin molecules made by a hybridoma cell derived from NS-1 myeloma which produces its own light chains. In the hybridoma, only the antigen-binding sites formed between light and heavy chains of spleen origin are active (●), whereas those involving light chains of NS-1 origin are inactive (○).

has been used that may circumvent the need for HAT-sensitive mutants. The principle is that *before* fusion, myeloma cells are treated with an irreversible metabolic inhibitor whose intracellular existence in active form is very short, e.g., diethypyrocarbonate; following fusion, only hybrid cells survive, since they are rescued by intact enzymes of the normal partner (ref. 38; and A. Williamson, personal communication).

Immunization protocol is an important variable aimed at increasing the frequency of antigen-specific cells in the tissue used for fusion. It needs to be optimized for any given system depending on, for example, the antigen (soluble or particulate, purified or crude mixture), whether the immunization is induced by a live infection, perhaps localized in a particular tissue, or by a controlled dose of a nonreplicating substance, whether a particular isotype, e.g., IgA or IgE is desired, etc. When the spleen of a preimmunized mouse is to be used, an intravenous challenge with the antigen three to four days before fusion gives consistently good results. In the production of human hybridomas, the most accessible source of immune cells is the peripheral blood, and *in vitro* stimulation of such cells prior to fusion may improve the chances of obtaining specific hybrids.

Screening assay identifies the cultures that contain the required antibody. It is quite easy (and cheap) to produce a large number of hybrids in the primary cultures after fusion, but the further development of a hybrid into a cloned stable cell line demands a substantial expenditure of time, labor, and money. It is therefore desirable that the assay detects antibodies directly suited to the purpose for which they are eventually to be used. Hybrids producing some specificities may be rare, and the chances of identifying them increase with the number of independent cultures that can be screened; therefore, the test should also permit screening of up to several hundred culture fluids within a few days. Primary antibody binding assays—e.g., indirect immunofluorescence assay (IFA), enzyme-linked immunoadsorbent assay (ELISA), or variants of radioimmunoassay (RIA)—have the common advantage of being rapid, and are suitable for initial screening in most situations. Furthermore, all three tests can identify any specified isotype, which in turn gives an indication of some functional properties of the antibody, e.g., its capacity to activate complement or to bind to certain effector cells. IFA has the additional advantage that it yields rapid preliminary information about the localization and distribution of antigens within tissue, cell, or developmental cycle, and it has been the assay of choice in most laboratories that raise monoclonal antibodies to parasites.

None of the binding assays is very informative as to the actual biological activities of the antibodies, which may be intended for *in vitro* or *in vivo* functional studies. In some instances, it may be possible to use a complement-dependent cytotoxicity assay (9), but most functional assays for antiparasite antibody are too cumbersome for rapid testing of many samples or require much more concentrated antibody than can be obtained in culture fluids (7,20,21,27,29,30). Under these circumstances, the only possible strategy is to choose for the initial screening a binding assay likely to provide most information about those features of both the antibody and the antigen that are expected to be of importance for the functional studies (e.g., antibody isotype, surface, and/or stage-specific antigen) and to develop the more promising hybridomas to the point where high-titer ascitic fluids or sera are ready for further screening by functional tests.

Cloning removes unwanted hybrids and variants that produce irrelevant antibody (e.g., to host antigens) or no antibody and that may overgrow the producer cells. It is recommended that cloning be done as soon as possible after identification of a promising culture and that it be periodically repeated. In addition, recloning facilitates the isolation of occasional desirable variants that no longer produce the chains of the myeloma parent (in hybrids derived from a producer myeloma that may be making hybrid antibody molecules). Moreover, heavy-chain variants that may appear at the low frequency of 0.1 to $10/10^6$ cells in cloned hybrid populations (V. T. Oi and L. A. Herzenberg, personal communication) can be identified and isolated by a one-step procedure using a fluorescence activated cell sorter. This offers a very promising approach to building up a library of antibodies of identical specificity (V region) joined to different C_H regions and therefore with different biological properties.

POTENTIAL CONTRIBUTIONS OF MONOCLONAL ANTIPARASITE ANTIBODIES

Some of the possible uses are listed in Table 1, while examples of current applications are given in Table 2 and the following section of this chapter. The most immediate uses are concerned with identification, characterization, and purification of individual antigens from the complex antigenic mixtures represented by parasites and the contaminating host material (17,24,26). Second, monoclonal antibodies are already contributing to the identification of those parasite antigens and the effector immune mechanisms that are involved in protection. Thus, monoclonal antibodies to several stages of the developmental cycle of malaria parasites identify unambiguously a number of antigens of great potential value for the formulation of antimalaria vaccine (7,27,29,31). While most of the antigens are stage-specific to either sporozoites, merozoites, or gametes, their species and strain specificity is largely unknown. Some of these studies, as well as those on eosinophil-mediated damage to *Schistosoma mansoni* schistosomula in the presence of monoclonal rat IgG2a (34), allow very precise analysis of the role played by antibody in immune pro-

TABLE 1. *Potential contributions of monoclonal antiparasite antibodies*

Parasite Immunology
Definition of antigenic determinants, and immunochemical and biochemical characterization of the carrier antigen molecules
Immunoadsorbent purification of antigens for vaccination and other purposes, particularly valuable for rare or poorly immunogenic molecules
Antigenic diversity and antigenic variation
Identification of antigens that are likely to stimulate protective responses *in vivo*
In vitro studies on mechanisms of antibody-dependent immunity
Identification of antigens involved in immunopathological reactions
Internally radiolabeled antibody reagents (^{35}S, ^{3}H, ^{14}C)
Medicine
Immunodiagnosis: presence of antigen in infected host (high-avidity antibody for radioimmunoassays)
identification of the species or strains of pathogens
Epidemiology
Standardization of diagnostic reagents and vaccines
(Drug targeting)
(Immunotherapy)
Molecular biology
Probes to identify products synthesized in cell-free systems or in bacteria
Parasitology
Species-specific markers for taxonomy
Clone-specific markers for genetic studies
Identification of developmental stages or lineages

TABLE 2. *The existing monoclonal antibodies specific for parasites*

Parasite (stage)	Biological activity	Other application	Ref.
Schistosoma mansoni (schistosomula)	Complement-dependent cytotoxicity Eosinophil-mediated cytotoxicity		34
Toxoplasma gondii (tachyzoites)	Complement-dependent cytotoxicity		9, 22
Plasmodium berghei (sporozoites)	Protection *in vivo*		29
P. gallinaceum (gametes)	Inhibition of transmission		31
P. yoelii (merozoites)	Protection *in vivo*		7
P. falciparum (merozoites)	Growth inhibition *in vitro*		27
P. falciparum (asexual blood forms)		Antigenic diversity within species	[a]
Theileria parva (macroschizonts)		Antigenic diversity within species	28
Leishmania brazilliensis and *mexicana*		Antigenic differences among species	18
Trypanosoma sp.		Antigenic variation	17, 25, 26

[a]See text.

tection. Furthermore, it is a good omen for future vaccination to observe that antibody to a single antigenic determinant of a parasite may directly or indirectly initiate the interruption and final elimination of the infection. The third area of application of antiparasite monoclonal antibodies as typing and diagnostic reagents is also yielding promising results. Monoclonal antibodies specific for either *Leishmania mexicana* or *L. brazilliensis* differentiate between the causative pathogens of cutaneous and mucocutaneous leishmaniasis and thus allow an early diagnosis and treatment of patients likely to suffer from serious disfiguring disease (18). As the numbers of both available antibodies and laboratories that use them grow, most of the applications mentioned in Table 1 will become increasingly routine, whereas some applications, like drug targeting or immunotherapy, may remain hypothetical for a considerable time.

ANTIGENIC DIVERSITY OF THE HUMAN MALARIA PARASITE *PLASMODIUM FALCIPARUM*

In this last section, I would like to present some data obtained with monoclonal antimalaria antibodies in Edinburgh. The principal objective of the work has been to investigate the existence and the extent of antigenic polymorphism within the species *P. falciparum*.

The clinical manifestations of malaria are associated with the asexual erythrocytic cycle of the parasites (merozoite→ring→trophozoite→schizont→merozoites). While many components of the blood forms are recognized as antigenic, only a limited number of them are likely to be important as targets of protective immune responses, e.g., surface antigens of the extracellular merozoites and/or of the infected red cell membrane and merozoite product(s) involved in invasion. It is not known whether the postulated protective antigens of *P. falciparum* show diversity among different populations of the parasite, although such knowledge is obviously relevant to formulation of a vaccine. Furthermore, polymorphism of any antigen could be exploited as a source of markers additional to the existing isoenzyme variants (31a), e.g., for epidemiological studies.

Mouse monoclonal antibodies against the blood parasites of a Thai isolate (K1) were developed following the protocol of Perrin et al. (27). When tested by IFA on fixed thin blood films of asynchronous cultures of K1, the antibodies presented several distinct staining patterns. These reproducible patterns pointed to characteristically restricted distributions of the antigens within the erythrocytic cycle and/or within the parasite cells and have been used to classify the antibodies and the antigens defined by them into five groups:

1. Antibodies reactive with an antigen or antigens apparently located in the cytoplasm of all asexual blood forms and gametocytes (five antibodies).
2. Antibodies reactive strongly with merozoites both within and outside of schizonts and, more weakly, with rings but not with trophozoites; the staining of merozoites indicates that the antigen(s) may be associated with a large organelle that is not very well defined in rings (eight antibodies).

3. Antibodies reactive with all asexual blood forms but not with gametocytes; mature schizonts present a characteristic appearance reminiscent of raspberry fruit owing to staining of individual merozoites within; merozoites retain the antigen(s) after release from schizonts (three antibodies).
4. Antibodies reacting with schizonts and merozoites in a pattern very similar to that of group 3, but the antigen(s) is(are) more stage-specific, appearing around early schizogony (five antibodies).
5. Antibodies reactive with the more mature parasites but not with merozoites or rings; the antigen(s) may also be found as inclusions within the infected erythrocytes (two antibodies).

This panel of 23 monoclonal antibodies was used to type 18 isolates of *P. falciparum* of different geographical origins. Each sample tested by IFA included 10^4 schizonts and 10^5 to 10^6 other stages, and each isolate was tested at least three times using parasites harvested at weekly intervals. The reactions were scored visually, using as the main criterion the intensity of staining of *individual* parasites (at the appropriate stages of development) relative to staining of the homologous K1 isolate that was included on all occasions.

Seventeen of the antibodies from the panel react equally well with all isolates, indicating that the respective determinants are either ubiquitous or that variants are rare. All antibodies belonging to groups 1 and 2, and two antibodies out of three, one out of five, and one out of two in groups 3, 4, and 5, respectively, define these "common" specificities.

The six remaining monoclonals clearly differentiate between isolates (Fig. 4). Antibodies 6.1, 7.3, and 7.6 are similar to one another, staining parasites of any given isolate either well or not at all. This correlation among their reactivities on different isolates, as well as the similarity of their staining patterns (group 4), indicates that they probably react with the same antigen. The antigen has a worldwide distribution, although negative variants are common in Thailand and are also present in Africa. Antibody 7.1 (also group 4) gives subnormal dull rather than negative reactions on a number of isolates but its reactivities correlate well with those of the 6.1, 7.3, 7.6 trio. It is considered to define another specificity, perhaps on the same antigen. Antibodies 7.5 (group 3) and 5.1 (group 5) clearly define two additional specificities unrelated to the other. To summarize, determinants defined by 23 different monoclonal antibodies against *P. falciparum* have worldwide distribution. The majority of the determinants are common but at least four different sites show diversity, and parasites lacking the specificities occur both in Thailand and Africa. Three of the strain-specific determinants are expressed by merozoites. It may be added that owing to the existence of common determinants as well as to the independent distribution of the strain-specific ones among isolates, this kind of serological analysis would be almost impossible before the advent of monoclonal antibodies.

Serotyping with strain-specific monoclonal antibodies promises to become a valuable addition to the existing methods for identification of different populations

ISOLATE			SPECIFICITY					
Country	Area	Code	5·1	6·1	7·3	7·6	7·1	7·5
Thailand	Kanchanaburi	K 1	■	■	■	■	■	■
		K 28	■	□	□	□	⊡	⊡
		K 34	□	■	■	■	■	⊡
	Prabuthabath	PB1	□	□	□	□	⊡	⊡
	Songkhla	SK15	■	□	□	□	⊡	■
		SK16	■	■	■	■	■	■
		SK17	■	□	□	□	⊡	■
		SK18	■	□	□	□	⊡	⊡
	Sriracha	S 2	■	□	□	□	⊡	⊡
		S 118	■	■	■	■	■	■
	Tak	T 22	■	■	■	■	■	■
Burma		T 17	■	□	□	□	⊡	⊡
Sri Lanka		SL3	■	■	■	■	■	■
Indonesia		NK58	■	■	■	■	■	■
Gambia		BW	■	■	■	■	■	■
		G1	■	□	□	□	□	■
Uganda		PaloAlto	□	□	□	□	□	■
Honduras		M23	■	■	■	■	■	■

FIG. 4. Distribution of monoclonal-defined specificities in isolates of *P. falciparum*. Symbols: ■, positive bright; ⊡, positive dull; □, negative. See the text for details of the indirect immunofluorescence typing test.

TABLE 3. *Typing of* P. falciparum *cultures for isoenzyme and antigen markers*

	Isoenzyme[a]					
	GPI		PEPE		Antigenic specificity	
Culture	1	2	1	2	6.1	9.5
Isolate NF58	−	+	−	+	+	−
Clone 94	+	−	+	−	+	−
Clone 19	−	+	+	−	−	−
Isolate G1	−	+	+	−	−	+
Isolate Tak 17	+	+	+	−	−	+
Isolate Tak 20	+	−	+	−	− and +	− and +

[a]See ref. 39 for methods of enzyme typing. See text for the details of antigen typing.

of *P. falciparum*. The antigen markers are independent of, e.g., GPI and PEPE isoenzyme markers, and Table 3 illustrates how the two marker systems can complement each other. The isoenzymes differentiate between isolate NF58 and clone 94, which both have the same antigen markers. In contrast, clone 19 and isolate G1, identical with respect to the isoenzymes, can be distinguished by their antigen markers. In addition, heterogeneous mixed isolates can be identified with the help of either the isoenzymes (Tak 17 contains two GPI variants) or the antigen markers

(Tak 20 by the presence of organisms negative as well as of those positive for antigenic specificities defined by antibodies 6.1 and 9.5).

ACKNOWLEDGMENTS

The monoclonal antibodies were prepared in collaboration with Mrs. Gill Morgan (Department of Molecular Biology) and the typing of *P. falciparum* isolates was done in liaison with Dr. David Walliker (Department of Genetics). I would like to thank Professor G. H. Beale, Dr. John Scaife, and Dr. Spedding Micklem for continuous interest, stimulation, and encouragement throughout the work. The project has been supported by grants from The Wellcome Trust and from the Medical Research Council.

REFERENCES

1. Adams, J. M. (1980): *Immunol. Today*, 1:10.
2. Atassi, M. Z. (1975): *Immunochemistry*, 12:423.
3. Atassi, M. Z. (1978): *Immunochemistry*, 15:904.
4. Atassi, M. A., and Smith, J. A. (1978): *Immunochemistry*, 15:609.
5. Croce, C. M., Linnenbach, A., Hall, W., Steplewski, Z., and Koprowski, H. (1980): *Nature*, 288:488.
6. Fazekas de St. Groth, S., and Scheidegger, D. (1980): *J. Immunol. Methods*, 35:1–21.
7. Freeman, R. R., Trejdosiewicz, A. J., and Cross, G. A. (1980): *Nature*, 284:366.
8. Galfre, G., Milstein, C., and Wright, B. (1979): *Nature*, 277:131.
9. Handman, E., and Remington, J. S. (1980): *Immunology*, 40:579.
10. UNDP/World Bank/WHO (1980): *Hybridoma Technology with Special Reference to Parasitic Diseases*. Special Programme for Research and Training in Tropical Diseases.
11. Kabat, E. A. (1976): *Structural Concepts in Immunology and Immunochemistry*, 2nd ed. Holt, Rinehart and Winston, New York.
12. Kaufman, Y., Berke, G., and Eshhar, Z. (1981): *Proc. Natl. Acad. Sci. USA*, 78:2502.
13. Kearney, J. F., Radbuch, A., Liesegay, B., and Rajewsky, K. (1979): *J. Immunol.*, 123:1548.
14. Köhler, G., and Milstein, C. (1975): *Nature*, 256:495.
15. Köhler, G., and Milstein, C. (1976): *Eur. J. Immunol.*, 6:511.
16. Melchers, F., Potter, M., and Warner, N. L. (1978): *Curr. Top. Microbiol. Immunol.*, 81:1–240.
17. Lyon, J. A., Pratt, J. M., Travis, R. W., Doctor, B. P., and Olenick, J. G. (1981): *J. Immunol.*, 126:134.
18. McMahon Pratt, D., and David, J. R. (1981): *Nature*, 291:581.
19. Milstein, C., Clark, M. R., Galfre, G., and Cuello, A. C. (1980): In: *Prog. Immunol.*, 4:17–33.
20. Mitchell, G. F. (1979): *Immunology*, 38:209.
21. Wakelin, D. (1978): *Nature*, 273:617–620.
22. Naot, Y., and Remington, J. S. (1981): *J. Immunol. Methods*, 43:333.
23. Olsson, L., and Kaplan, H. S. (1980): *Proc. Natl. Acad. Sci. USA*, 77:5429.
24. Pearson, T., and Anderson, L. (1980): *Anal. Biochem.*, 101:377.
25. Pearson, T. W., Pinder, M., Roelants, G. E., Kar, S. K., Lundin, L. B., Mayor-Withey, K. S., and Hewett, R. S. (1980): *J. Immunol. Methods*, 34:141.
26. Pearson, T. W., Kar, S. K., McGuire, T. C., and Lundin, L. B. (1981): *J. Immunol.*, 126:823.
27. Perrin, L. H., Ramirez, E., Lambert, P. H., and Miescher, P. A. (1980): *Nature*, 289:366.
28. Pinder, M., and Hewett, R. S. (1980): *J. Immunol.*, 124:1000.
29. Potocnjak, P., Yoshida, N., Nussenzweig, R. S., and Nussenzweig, V. (1980): *J. Exp. Med.*, 151:1505.
30. Cohen, S. (1980): *Prog. Immunol.*, 4:763–781.
31. Renner, J., Carter, R., Rosenberg, Y., and Miller, L. H. (1980): *Proc. Natl. Acad. Sci. USA*, 77:6797.
31a. Sanderson, A., Walliker, D., and Molez, J. F. (1981): *Trans. R. Soc. Trop. Med. Hyg.*, 75:263.
32. Schulman, M., Wilde, D. C., and Köhler, G. (1978): *Nature*, 276:269.

33. Secher, D. S. (1980): *Immunol. Today*, 1:22–26.
34. Werwaerde, G., Grzych, J. M., Bazin, H., Capron, M., and Capron, A. (1979): *C.R. Acad. Sci. Ser. D*, 289:725.
35. Wiley, D. C., Wilson, I. A., and Skehel, J. J. (1981): *Nature*, 289:373.
36. Williamson, A. (1976): *Annu. Rev. Biochem.*, 45:467.
37. Winkelhake, J. L. (1978): *Immunochemistry*, 15:695.
38. Wright, W. E. (1978): *Exp. Cell Res.*, 112:395.

Molecular Biology of Parasites, edited by
J. Guardiola, L. Luzzatto, and W. Trager.
Raven Press, New York © 1983.

Regulation of Expression of Structural Genes in Eukaryotes: Transcriptional and Posttranscriptional Control Mechanisms

A. Cascino

International Institute of Genetics and Biophysics, 80125 Naples, Italy

A basic property of cells is their ability to regulate the intracellular concentration of specific gene products either in response to extracellular signals or by switching on and off individual members of a gene family, as in the case of the globin genes, at different times during development. The expression of at least some eukaryotic genes has been proved unequivocally to be controlled at the level of transcription.

Detailed studies on regulatory mechanisms in prokaryotes and related bacteriophages has revealed that gene regulation takes place basically at the level of transcription, involving direct and indirect interaction between RNA polymerase and signal DNA sequences. Also, in the case of higher organisms, the expression of at least some eukaryotic genes—ovalbumine, for instance—has been proved to be controlled at the level of transcription. However, the precise nature of the signal sequence of eukaryotic genes is not yet elucidated in enough detail to enable us to formulate precise models of transcription regulation.

This chapter deals only with transcription initiation and the processing of precursor mRNA molecules of eukaryotic interrupted genes coding for structural proteins. These genes are transcribed only by RNA polymerase B (Chambon, 1975).

PROMOTER DEFINITION IN PROKARYOTES

Basic research on the regulation of transcription in prokaryotes has shown that regulation of transcription is achieved by modulating the efficiency of recognition and binding of RNA polymerase to specific DNA sequences that specify starting (promoters) and termination sites (attenuators) of RNA synthesis. The particular base sequence of the DNA with which the polymerase must interact in large part determines the strength of the signal.

The comparison of the sequences of several promoters of prokaryotic genes has shown that (a) all of them are located in regions of the DNA 5′ end with respect to the structural genes; (b) a 5′-TATAATG-3′ related sequence, Pribnow box (Pribnow, 1975), is located approximately 10 base pairs "upstream" from the mRNA starting points; (c) at about position -35, a second region of homology (5′-TTG-

3′) is found; and (d) the most frequent starting nucleotide of mRNA is adenosine (Fig. 1) (for a review, see Rosemberg and Court, 1980).

EUKARYOTIC RNA POLYMERASES

In eukaryotes, three different RNA polymerases exist that catalyze the synthesis of ribosomal RNA (enzyme A), mRNA (enzyme B), and 5S RNA and transfer RNA (enzyme C), respectively. This chapter deals only with transcription initiation and precursor mRNA (pmRNA) splicing of structural genes transcribed by RNA polymerase B.

ANATOMY OF EUKARYOTIC STRUCTURAL GENES

5′ Region

Restriction analysis and molecular cloning techniques have revealed that the starting point for transcription of structural eukaryotic genes corresponds to the base coding for the 5′ terminal nucleotide of the pmRNA. In analogy to prokaryotic genes, one should then expect to locate eukaryotic promoter sequences upstream to the 5′ end of the transcription unit.

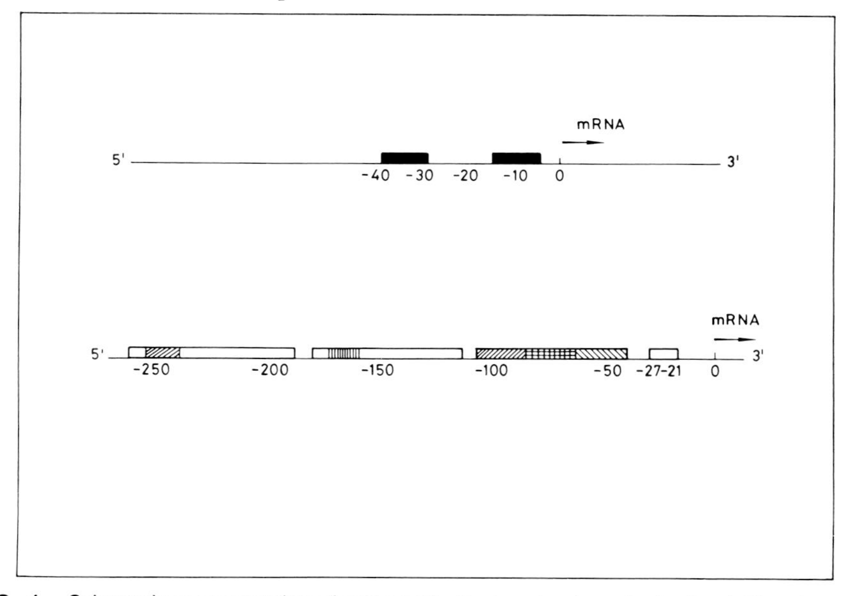

FIG. 1. Schematic representation of prokaryotic (top) and eukaryotic (bottom) 5′-region sequences flanking structural genes. **Top:** Prokaryote promoter: filled boxes indicate two highly conserved regions. **Bottom:** The 5′ region of eukaryotic structural gene: the two 72-base-pair repeats (I and II) and the TATA box are shown by boxes. The five GC-rich blocks are shown in cross-hatched boxes. All sequences refer to noncoding DNA. Numbering system refers to 0 as the first nucleotide of the transcript.

Comparison of the sequences upstream from starting point of transcription of several eukaryotic genes has, in fact, revealed the existence of some homology regions (Fig. 1): (1) an AT-rich region centered about 25 base pairs upstream from the mRNA starting site, also known as the "TATA box," a sequence that might be the equivalent of the Pribnow box of prokaryotes; (2) a $5'$-GG$^{C}_{T}$CAATCT-$3'$ sequence has been found at position -70 to -80; (3) the mRNA starting site appears to be an adenosine residue, surrounded by pyrimidines; and (4) a tandem repeat sequence, containing a GG-rich block, is found more than 150 base pairs upstream of the mRNA starting site. This sequence is indispensable for *in vivo* transcription.

These homologous sequences are not found upstream from the starting points of genes transcribed by RNA polymerases A and C.

Interrupted Genes

A second feature of eukaryotic genes that might be involved in the regulation of gene expression (posttranscriptional regulation) involves the discovery of spliced mRNA and the finding that several genes consist of noncontiguous blocks of coding sequences separated by intervening sequences (Fig. 2) (for a short review, see Lewin, 1980). It has been demonstrated that nuclear pmRNA—to which, in most cases, a polyadenylated tail is added—is shortened by the removal of internal RNA sequences and only mature mRNAs are exported in the cytoplasm. In most cellular

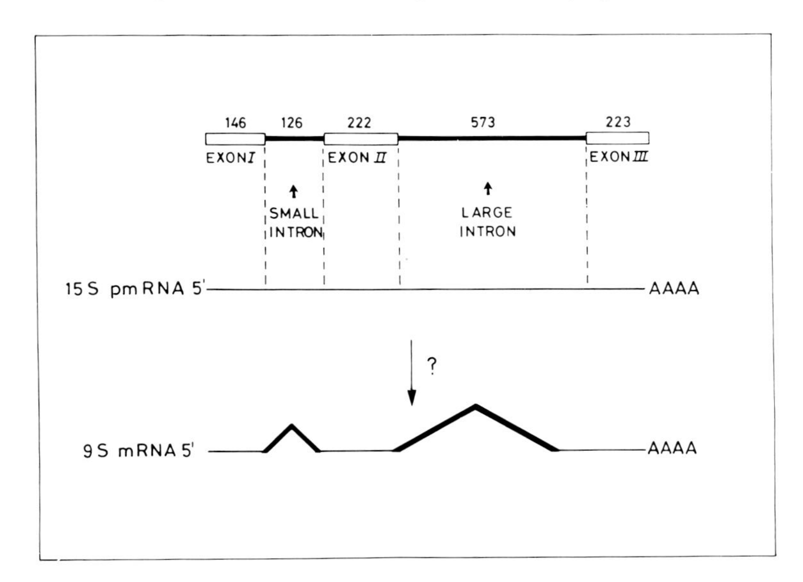

FIG. 2. Scheme of rabbit β-globin pmRNAs. The β-globin gene is constituted by three coding sequences (exons) separated by two intervening sequences (introns). The numbers represent the nucleotide length of each block. The bold peaks (⋀) indicate the intron sequences present in the 15S pmRNA molecule that are removed to produce 10S mRNA.

genes, the splicing pathway appears to be invariant: the same pmRNA always gives rise to the same mature mRNA by removal of defined introns. Important exceptions exist in which alternative processing patterns are available to a single gene. They give rise, upon translation, to proteins whose amino acid sequences are partially in common and partially different. Furthermore, even in cases where splicing always produces invariant mRNA, removal of the intron seems to involve not simply the recognition of the two 5′ and 3′ proximal and distal sites adjacent to the exons. It has been found, in fact, that sequences internal to introns are spliced and that removal of one portion of the intron does not interfere with the subsequent elimination of the rest of the intron.

It has not been shown, however, whether these intermediates are necessary steps for the removal of intron or whether they represent dead ends in the splicing pathways.

PROMOTER SEQUENCES IN THE TRANSCRIPTION INITIATION OF EUKARYOTIC GENES TRANSCRIBED BY RNA POLYMERASE B

In Vitro Analysis

Point Mutations

The techniques used to investigate eukaryotic promoter sequences are based on restriction-enzyme, molecular-cloning, and DNA-sequencing methods. *In vitro* analysis is usually performed using the appropriate test DNA template (previously cut at a known map position in order to obtain RNA species of discrete size, which are produced by runoff termination), depending on where specific initiation occurs. The length of the radioactively labeled runoff transcript is determined by electrophoresis in a sequencing gel followed by autoradiography and comparison with the position of the 5′ end of the *in vivo* transcript. Using *in vitro* systems, S100 KB cell extract or injection into oocyte nuclei, and labeling the RNA synthesized *in vitro* with radioactive nucleoside triphosphates, Waslylyk et al. (1980) demonstrated that *in vitro* transcription of the conalbumine TAGA box mutant is about 5% of that of wild-type TATA sequence template. Therefore, the single base-pair transversion (T to G) at position −29 in the TATA box drastically decreases the efficiency of specific transcription of conalbumine DNA.

This finding supports the hypothesis that the TATA box sequence is an important element for initiation of specific *in vitro* transcription. Since a promoter sequence is defined as a DNA sequence that is recognized by the RNA polymerase and to which RNA polymerase binds, these findings do not, however, prove that the TATA box is indeed part of the eukaryotic promoter sequence. In fact, in a quite crude *in vitro* system there is no experimental evidence that the DNA template is in excess over essential factor(s) present in the S100 extract. Analogously, no data exist on direct interaction between the TATA box region and RNA polymerase B and/or factor present in the system.

Deletion Mutants

Specific RNA polymerase B initiation is drastically reduced in mutants with deletions involving the TATA box sequence. When the TATA box sequence is present in the template, RNA polymerase B begins transcription about 30 base pairs downstream. In its absence, the enzyme initiates at multiple sites over a broad region. Mathis and Chambon (1981) propose that the TATA box functions *in vitro* by fixing initiation of transcription within a narrow region, whereas the tandem repeat sequence, located more than 150 base pairs upstream (Fig. 1) and indispensable for *in vivo* transcription, is not essential *in vitro*.

The interpretation of the results obtained both *in vivo* and *in vitro* with deletion mutants is limited, since a deletion can be more than the simple absence of a particular sequence. In fact, new DNA sequences are fused to the deletion end points. This makes it impossible to rule out the suggestion that the observed effects are due to an effect of the replacing sequences rather than to direct alternation of the promoter region.

In Vivo Analysis

Benoist and Chambon (1981) have shown that deletion of all sequences upstream from the cap site (Fig. 1) abolishes *in vivo* expression of SV40 early genes in cloned rat fibroblast cell lines, transformed with the appropriate SV40 deletion mutant DNA or in injected *Xenopus* oocytes. The removal of the TATA box sequence results in the generation of multiple cap sites up to 270 base pairs downstream from the normal 5′ end. This finding is substantiated by the fact that polyoma late genes and adenovirus DNA binding protein gene exhibit a little heterogeneity of mRNA 5′ end. These genes do not have TATA box sequences at the expected position.

SV40 early gene expression is drastically reduced *in vivo* if deletions far upstream from the TATA box are used, strongly suggesting that the essential sequence is in the region of the 72-base-pair tandem repeats (Fig. 1).

Taking into account the results reported above, a preliminary model for the specificity of transcription initiation could be as follows: the TATA box sequence specifies the initiation of transcription 25 base pairs downstream and the tandem 100 base pairs upstream repeat sequences are also necessary. The last might be the binding site of the protein(s) required for transcription or might be involved in the generation of an "open" chromatin structure that makes the sequence downstream accessible to RNA polymerase B.

Since nicked DNA templates are used in the *in vitro* systems and no chromatinlike structures are reconstituted, the *in vitro* promoter site for RNA polymerase B would consist only of the TATA box and of the −70 to −80 repeated sequences. The *in vivo* requirement for the far-upstream sequences, which do not correspond to the structure of prokaryotic promoter sites, might reflect the role of chromatin structure in the control of eukaryotic gene expression.

Further support is provided by the observation that actively transcribed genes are more sensitive to deoxyribonuclease I digestion than are nontranscribed sequences:

DNase I sensitivity of active chicken globin genes extends for many kilobase pairs on either side (Stalder et al., 1980).

POSTTRANSCRIPTIONAL REGULATION

Precursor mRNA Processing and Export into Cytoplasm

Although very little experimental data exist, as an example of RNA splicing and its possible relationship to regulatory mechanisms, we will describe what is known about the globin gene family.

These genes are expressed at different times of development, and mutations are known that affect the expression of α- and β-globin in humans (α- and β-thalassemia) or the switch from fetal to adult globin gene expression.

In four different β-thalassemia patients, Kantor et al. (1980) showed that although the rate of pmRNA transcription is quite normal with respect to wild type, abnormal RNA species accumulate. Their data suggest that mutations that affect RNA processing might be the cause of some β-thalassemia.

The thalassemias are inherited disorders of human hemoglobin synthesis characterized by defective production of the α- and β-globin components of normal adult hemoglobin. In homozygous β°-thalassemia, no β-globin is synthesized, although many such patients have detectable amounts of β-globin mRNA sequences. In homozygous β^{+}-thalassemia, there is only a quantitative reduction in the amount of mature β-globin mRNA. Fine restriction mapping of several patients with β°- and β^{+}-thalassemia shows that the general organization of the β-globin gene is normal. Several reports also indicate that the decreased accumulation of β-globin mRNA in β^{+}-thalassemia is probably not due to mutations affecting transcription rate, but rather to an abnormal pmRNA processing or instability of mRNA.

The pmRNA β-Globin Molecule

The initial product of transcription of the human β-globin gene is an 1,800-base pair polyadenylated RNA molecule that is present in the nucleus of erythroid cells. The pmRNA is then processed to mature 10S mRNA, which accumulates in the cytoplasm and is translated there.

Since β-globin gene in many different species contains two introns, a small intron of about 100 bases close to the 5′ end of the gene and a larger one of about 850 bases toward the 3′ end of the gene, defective pmRNA processing might be the molecular basis for some of the β^{+}-thalassemias.

Splicing defects can lead either to RNA molecules that cannot be translated or to a decreased amount of the correct mRNA as a result of competition with the newly introduced splicing site. Splicing defects can then represent regulatory mutations, since they cause a quantitative reduction in the intracellular concentration of a structurally normal product.

β-Globin mRNA Processing in β^+-Thalassemia

Kantor et al. (1980) used hybridization analysis of labeled RNA extracted from bone marrow cells to show that four different β^+-thalassemic patients had a higher concentration of β-globin mRNA precursors. In two who had been studied in detail, the cytoplasm contained lower amounts of β-globin mRNA than did controls. One of these patients showed high levels of an abnormal 650-nucleotide-long molecule containing sequences transcribed from both the large intron and the part of the coding region of the globin gene. In the other, a 1,320-nucleotide-long molecule from which only part of the large intron sequence had been removed was shown to accumulate. These alterations in processing could then result in a delay of the transport of β-globin mRNA into the cytoplasm.

Removal of Both Intervening Sequences of Rabbit β-Globin pmRNA Is Highly Complex

Using a more sensitive technique (S1 nuclease digestion of the hybrids between total bone marrow RNA and the appropriate denatured rabbit β-globin gene DNA fragments labeled with ^{32}P, followed by alkaline electrophoresis of the undigested fragment, blotting on nitrocellulose filter paper, and autoradiography), Grosveld et al. (1981) were able to characterize the transcripts of the rabbit β-globin gene. Basically they found a number of partially spliced RNAs that are probably intermediates in the pathway from the precursor to the mature β-globin mRNA.

Their findings were as follows: (1) the largest intermediate RNA lost only about 40 bases of the 126 nucleotides of the small intron, (2) the second major intermediate lost the entire small intron, and (3) the third intermediate lost all but approximately 90 nucleotides of the large intervening sequence.

How Single-Base Substitution Might Affect pmRNA Maturation

The analysis of several spliced genes has revealed that in all cases the exon-intron and intron-exon boundaries contain the two dinucleotides GT and AG, respectively, at unique positions. This feature is obviously insufficient to identify the splicing points on pmRNA molecules. Splicing points can be assigned tentatively by analysis of a quite large number of nucleotides flanking the intron boundaries with the use of a computer program that can show the most probable splicing points of any given sequence. Splicing specificity should ultimately reside in the nucleotide sequence of the primary transcript, since, first, protein coding gene transcripts are correctly spliced in heterologous organisms and, second, correct splicing also takes place when the internal parts of the introns are deleted. Then, sufficient information to define a functional splice site should be encoded in a 20- to 30-base sequence.

A computer analysis applied to pmRNA sequences (Staden and Brownlee, *personal communication*) can identify several splicing consensus sequences. This identification is obtained by maximizing the homology of the two short nucleotide sequences at the exon–intron and intron–exon junctions. Consensus sequences are found to be located throughout the entire pmRNA sequences. Presumably, only a limited number of them are used for the correct maturation of a given molecule. Some of the others could become effective splicing points by single-base substitution.

Single-base substitution might then either remove an active splicing signal or introduce new splicing sites. This would interfere or compete with normal pmRNA processing, ultimately resulting in a relative decrease of the concentration of mature mRNA.

In this way, the genetic defect causing β-thalassemia may be, in part, an RNA-processing disease. It is not yet possible, however, to estimate the incidence of aberrant splicing events in the naturally occurring, observable mutations of an interrupted gene.

REFERENCES

1. Benoist, C., and Chambon, P. (1981): *Nature*, 240:304–310.
2. Chambon, P. (1975): *Ann. Res. Biochem.*, 44:613–638.
3. Grosveld, G. C., Koster, A., and Flavell, A. (1981): *Cell*, 23:573–584.
4. Kantor, J. A., Turner, P. H., and Nienhuis, A. W. (1980): *Cell*, 21:149–157.
5. Lewin, B. (1980): *Cell*, 22:324–326.
6. Mathis, D. J., and Chambon, P. (1981): *Nature*, 290:310–315.
7. Pribnow, D. J. (1975): *J. Mol. Biol.*, 99:419–443.
8. Rosemberg, M., and Court, D. (1980): *Annu. Rev. Genet.*
9. Stalder, J., Larsen, A., Engel, J. D., Dolan, M., Groudine, M., and Weintraub, H. (1980): *Cell*, 20:451–460.
10. Wasylyk, B., Derbyshire, R., Guy, A., Molko, D., Roget, A., Teoule, R., and Chambon, P. (1980): *Proc. Natl. Acad. Sci. USA*, 77:7024–7028.

Molecular Biology of Parasites, edited by
J. Guardiola, L. Luzzatto, and W. Trager.
Raven Press, New York © 1983.

In vitro Translation Systems

Anna Maria Guerrini and Maurizio Iaccarino

International Institute of Genetics and Biophysics, CNR, 80125 Naples, Italy

The *in vitro* translation systems are cell-free systems with intact machinery for protein synthesis. They have been developed to study many problems of molecular biology, and in particular, they have been used to understand the complex events of transcription and translation, tRNA function, mRNA processing, and so on. This has been achieved, since these systems allow us to fractionate the required factors and to analyze their role precisely. Moreover, these systems represent a situation very similar to that found in living cells, since it is possible to study the processing of specific macromolecules added to the system. The *in vitro* translation systems are bacterial extracts (prokaryotic systems) or lysates from mammalian and plant cells (eukaryotic systems). They show several differences, some of which are inherent to the biology of the cells themselves. The most important is that the prokaryotic systems use DNA as a source of mRNA, whereas the eukaryotic ones use mRNA. Although many of the studies described in this volume concern eukaryotic parasites, we describe in detail prokaryotic *in vitro* translation systems, as well as the eukaryotic ones, because they are of historical importance, and in a few respects, they are still better understood than the eukaryotic systems. Furthermore, an examination of the differences between them will lead to a better understanding of both. Chloroplast and mitochondrial systems will be described, as well, to give a complete view of the field. These organelles of eukaryotic cells have a translational machinery very similar to that of prokaryotic cells.

Since in the context of this volume this chapter deals with methodology, we have given preference to literature references describing methods in detail.

PROKARYOTIC SYSTEMS

The composition of the prokaryotic *in vitro* translation system is complex, and the reason for the use of some components is not obvious. We therefore give here a brief historical overview of the origin of this system.

Hoagland et al. (41) in 1958 showed that cell-free protein synthesis requires ribosomes, ATP, GTP, and tRNA. Later, Nirenberg and his collaborators (see, for example, ref. 67) gave the first evidence for the requirement of mRNA in protein synthesis by demonstrating that the addition of polyuridylic acid to the "ribosome fraction" stimulates polyphenylalanine synthesis. Because of this finding it was concluded that in the absence of added mRNA, amino acid incorporation is directed by preformed and/or newly synthesized endogenous mRNA (66). Preincubation of an extract, called S-30 extract, prepared from *Escherichia coli* cells with DNAse, prevented new mRNA synthesis, whereas preformed mRNA was destroyed. The S-30 extract catalyzed the incorporation of different amino acids when added with different ribonucleotide polymers ("synthetic messengers"), and this use of the S-30 extract was essential in the elucidation of the genetic code.

At the same time, the first complete protein, the coat protein of the bacteriophage f2, was synthesized in a cell-free extract of *E. coli* (64). (This is, however, a special case of *in vitro* translation, since the RNA is purified from an RNA phage: bacterial mRNAs are easily degraded during isolation, and moreover, they are probably more efficiently translated while they are synthesized. Instead, for complete protein synthesis, an intact mRNA should be translated at least once before it is degraded. The success with f2 messenger is at least in part due to its availability in pure form, but it is probably also due to a special secondary structure that renders it resistant to nucleases and efficient in attaching to ribosomes to initiate translation.) Shortly thereafter, the other two proteins encoded on the same messenger, maturation protein and RNA replicase, were also shown to be synthesized *in vitro* (see ref. 47 for a review). The coat protein was identified by the correspondence of the tryptic peptides of the product with the tryptic peptides of the protein prepared from phage particles. The *in vitro* and *in vivo* products showed the same behavior on Sephadex columns and during electrophoresis in polyacrylamide gels. When mRNA contained an amber mutation, the protein was terminated prematurely unless transfer RNA from a suppressor strain was used. Similar criteria were used to identify the RNA replicase and maturation protein synthesized *in vitro*.

These results were also useful for the study of some aspects of the mechanism of protein synthesis, such as initiation (1,90), elongation (42), and termination (14). Gene expression in RNA phages is regulated principally at the level of translation, and it appears to take place through variations in the secondary structure of the mRNA. The coat protein binds mRNA *in vitro* and inhibits translation of RNA (87), confirming an *in vivo* observation. Several other *in vivo* observations are confirmed with the *in vitro* translation system (47).

A different approach to the *in vitro* synthesis of proteins is the measurement of their enzymatic activity, which is a stringent requirement for fidelity and completeness of the synthesis. Salser et al. (79) reported the *in vitro* synthesis of active lysozyme coded by messenger extracted from *E. coli* cells infected with T4 phage. This system was relatively easy to develop for a number of reasons: (a) uninfected *E. coli* cells do not contain lysozyme (and therefore the background is very low); (b) the concentration of specific mRNA present in infected cells is high; and (c) the assay is very sensitive, and the protein is a single, short polypeptide chain (molecular weight of 18,000). These authors demonstrated that net synthesis of lysozyme was dependent on the mRNA added and was inhibited by chloramphenicol, puromycin, and ribonuclease.

In another study, *in vitro* synthesis of *E. coli* alkaline phosphatase was achieved by using mRNA extracted from derepressed cells (23). The assay in this case was based on the ability of newly synthesized radioactive alkaline phosphatase monomers to form dimers with excess unlabeled monomers. The dimers were then purified and their specific radioactivity measured.

Most of the *in vitro* translation studies in prokaryotes use DNA as a source of mRNA for translation. RNA polymerase is present in most S-30 extracts, and it is therefore added only to check saturation. Wood and Berg (92) showed that the DNA of several phages stimulates amino acid incorporation. Addition of deoxyribonuclease is inhibitory. Heat-denatured T2 DNA and single-stranded ϕX174 DNA produce no stimulation (93), suggesting that a double-helical DNA secondary structure is required for activity. Renatured T2 DNA or double-stranded ϕX174 DNA are in fact active. This *in vitro* system had different efficiencies with different DNAs, and it was improved (50) to obtain a system showing comparable efficiencies with different DNAs.

The use of DNA has several advantages: (a) RNA is continuously synthesized while translated; (b) it is likely that a nascent RNA chain is more actively translated than a finished one; (c) many DNA vectors are available containing the gene to be transcribed and translated *in vitro*, thus approaching the purity obtained with RNA phages.

The best studied system of DNA-directed RNA translation is the β-galactosidase system of *E. coli* developed by Zubay and his collaborators (62,99–101). The composition of this system is shown in Table 1. It was carefully studied by first optimizing the incorporation of [^{14}C]leucine into hot trichloroacetic-insoluble material (51), then optimizing for the synthesis of the α-peptide of β-galactosidase, a small part of the entire protein that—incubated with the remaining, inactive portion—gives activity (98), and finally optimizing for the synthesis of the entire protein (51). This system was essential to confirm and extend observations previously made *in vivo* on the regulation of expression of the *lac* genes, like the role of the repressor (98), cAMP (15), and ppGpp (75).

If the *lac* genes are fused to different promoters present on a specialized transducing phage the *in vitro* synthesis of β-galactosidase may be used to study regulation of expression of specific promoters, like, for example, *trp* (102) and *ilv* (89).

Besides β-galactosidase, other enzyme activities have been synthesized *in vitro* from specific DNAs, like L-ribulokinase from *ara* DNA (96), α- and β-glucosyl transferases from T4 phage DNA (33), *S*-adenosylmethionine-cleaving enzyme from T3 phage DNA (25), acetylornithinase from *arg* DNA (97), histidinol dehydrogenase from *his* DNA (75), galactokinase from *gal* DNA (91), and anthranilate synthetase and tryptophan synthetase from *trp* DNA (74).

The protocol used to prepare the S-30 extract from *E. coli* may be used to prepare an extract from *Klebsiella pneumoniae* (this laboratory, unpublished experiments). A similar protocol may be used to prepare the S-30 extract from *Bacillus subtilis* (53) or *Tetrahymena pyriformis* (19). An extensive effort is being made by Weissbach and collaborators to fractionate and purify the components of the S-30 extract (see, for example, 34, 48, 49). Using a fractionated extract, Jacobs et al. (44)

TABLE 1. *Components of the prokaryotic* in vitro *translation system*

Component	Amount[a]
Tris acetate, pH 8.2	44 μmol
Dithiothreitol	1.4 μmol
Potassium acetate	55 μmol
Each of the twenty L-amino acids[b]	0.22 μmol
Each of CTP, GTP, UTP, pH 7.5–8[c]	0.55 μmol
ATP, pH 7.5–8	2.2 μmol
Ammonium acetate	27 μmol
tRNA, *E. coli*	100 μg
Pyridoxine HCl	27 μg
NADP	27 μg
FAD	27 μg
Folinic acid	27 μg
p-Aminobenzoic acid	11 μg
Polyethylene glycol 6000	15 mg
Calcium acetate[d]	7.5 μmol
Magnesium acetate[e]	~15 μmol
Sodium phosphoenol pyruvate[f]	34 mg
3′,5′-cyclic AMP[g]	1 μmol
lac DNA	50 μg
S-30 protein[h]	6500 μg

[a]Indicated units per milliliter of incubation mixture.

[b]Stock solutions: 10 m*M* tyrosine at pH ~10, 400 m*M* cysteine (discard when a precipitate is formed), and 50 m*M* for each of the other 18 amino acids.

[c]GTP is needed for initiation of transcription at several promoters. The concentration of triphosphate needed for initiation of transcription is higher (150 μ*M*) than that required for elongation (15 μ*M*) (65). Since GTP is also used in protein synthesis and, moreover, it is unstable, change the GTP solution immediately when the system fails to work.

[d]Calcium acetate is used because it inhibits ribonuclease activity (43). Since calcium ions also inhibit translation initiation, it is useful to check with each S-30 extract if there is inhibition or activation. Alternatively, the S-30 extract should be prepared from a kasugamycin-resistant strain in which translation is resistant to inhibition by calcium.

[e]The optimum magnesium concentration should be checked with each S-30 extract, and it is very sharp. The S-30 extract contains magnesium, and further addition is required to reach the optimum (29, 30).

[f]This should be a freshly prepared solution of the trisodium salt from Calbiochem. Most of this substance is needed not to regenerate ATP but as a chelating agent. Addition of spermidine modifies the magnesium and phosphoenolpyruvate requirement (30).

[g]Specifically needed to activate the wild-type *lac* promoter, in conjunction with the cAMP binding protein present in the S-30 extract.

[h]The strain used to prepare the S-30 should be devoid of β-galactosidase: with the strains originated in Zubay's laboratory (used in most studies), the background of enzyme activity synthesized is negligible.

showed that transcription of the *lac* genes is stimulated four- to sixfold by the addition of ribosomes, i.e., by translation.

Although the *de novo* synthesis of enzymatic activity is a good criterion of fidelity of transcription-translation, it is pertinent to ask if the *in vivo* and *in vitro* products are identical. Webster et al. (90) studied the *in vitro* synthesis of the coat protein of phage f2 and found that although the protein isolated from phage particles has alanine with a free amino group at its terminus, the *in vitro* synthesized protein has the amino group of alanine bound to *N*-formylmethionine. More recently, on the other hand, it was shown (26) that the *in vitro* and *in vivo* amino terminal amino acid sequences of the *lacY* gene products are identical.

MITOCHONDRIA AND CHLOROPLASTS

Mitochondria and chloroplasts are capable of autonomous protein synthesis since they contain the entire equipment for the storage and expression of genetic information (80). These organelles should be considered separately from both prokaryotes and eukaryotes, since their DNA, RNA, ribosomes, and protein composition show quite particular features. Furthermore, the control of the macromolecular synthesis of the cellular organelles is exerted at two different intracellular levels: the cytoplasm and the organelle itself (5).

Protein synthesis in mitochondria and chloroplasts has been studied for several years by following the incorporation of radioactive amino acids into purified organelles and analyzing the effect of various drugs such as cycloheximide, antibacterial antibiotics, and inhibitors of transcription (31,36,55,76). The results of these studies gave relatively limited information about the synthesis of specific polypeptides but firmly demonstrated that mitochondrial and chloroplast DNAs code for specific translation products by a process very similar to, although distinct from, the prokaryotic protein-synthesizing system (80). The major part of *in vitro* experiments performed on mitochondria and chloroplasts using heterologous systems was carried out with the purpose of localizing the structural genes on mitochondrial and chloroplast DNA. Only recently the same *in vitro* experimental procedures have been used for the understanding of other biological processes such as RNA processing, tRNA function, effect of inhibitors, ionic conditions, etc. Since mitochondria and chloroplasts are quite different, especially with respect to the genetic code and decoding (9,28,57), we will describe the two systems separately.

Mitochondria

The first attempt to transcribe *mit* DNA *in vitro* and to translate the resulting mRNA in a reconstituted translation system was made by Blossey and Kuntzel (8). They reported that *E. coli* RNA polymerase efficiently transcribes *mit* DNA from *Neurospora crassa* and that the transcripts are translated in a submitochondrial *Neurospora* system, although the molecular weight of the products is strikingly different from that of the polypeptides synthesized in *Neurospora* mitochondria *in vivo*. They also demonstrated that the transcription-dependent protein synthesis was stimulated by addition of the formyl donor *N*-formyltetrahydropholic acid, which suggested that mitochondrial ribosomes use a bacteria-like chain-initiation mechanism. Shortly thereafter, Chuang and Weissbach (16) used a completely heterologous transcription-translation system prepared from *E. coli* extracts according to

Zubay et al. (100,101) programmed with rat liver *mit* DNA, but in this case also the pattern of labeled products obtained *in vitro* was significantly different from that obtained from *in vivo* labeled mitochondria.

A different approach was undertaken in the following years since (a) it was demonstrated that protein synthesis occurs on mitochondrial polysomes; (b) poly(A)-containing RNA, which hybridizes to *mit* DNA, is found in mitochondria (18,73); and (c) a number of hydrophobic polypeptides had been identified *in vivo* as products of the mitochondrial protein synthesis, i.e., the three subunits of the cytochrome *c* oxidase, the apoprotein of cytochrome *b*, two subunits of the oligomycin-sensitive ATPase, and finally, a small polypeptide (var 1) associated with the small subunit of the mitochondrial ribosomes (10,80).

Padmanaban et al. (68) obtained for the first time the *in vitro* synthesis of three cytochrome oxidase peptides by using a poly(A)-containing RNA fraction isolated from yeast mitochondria. They used a combination of the methods of Lindberg et al. (54) and Firtel et al. (27) to purify by poly(U) Sepharose chromatography a poly(A) RNA fraction that hybridized to purified *mit* DNA but not to either *E. coli* DNA or to nuclear yeast DNA. The RNA was tested in a S-30 *E. coli* system by following the [^{3}H]leucine incorporation into proteins precipitable by trichloroacetic acid. The incorporation was stimulated about fourfold over the endogenous levels upon addition of poly(A) RNA. To isolate the cytochrome oxidase peptides formed they treated the incubation mixture with antibodies to cytochrome oxidase and subjected the immunoprecipitates to sodium dodecyl sulfate (SDS)-polyacrylamide gel electrophoresis. Although their electrophoretic mobility was similar to that of the mitochondrially synthesized peptides, the results were not conclusive, since a possible nonspecific immunoprecipitation could have occurred in the reaction with the carrier cytochrome oxidase.

The same authors (40) confirmed these results with more sophisticated procedures and also demonstrated that translation of mitochondrial RNA fractions was stimulated by addition of Ca^{2+} to the S-30 system, the optimum being at 6 m*M* Ca^{2+}. The differential effect of Ca^{2+} on the *in vitro* translation of mRNAs from several sources was also studied by Halbreich (35), who reported that the *in vitro* translation of yeast *mit* RNA was less efficient than that of the RNA from MS2 or Qβ phages or the mRNA from T5 phage, but it could be enhanced several fold by addition of Ca^{2+}, whereas this addition severely inhibited the translation of viral RNAs (see above for the mechanism of action of Ca^{2+}).

The inability of heterologous systems to synthesize full-length mitochondrial proteins has become clear after the development of the DNA-sequencing methodology. The analysis of some yeast gene sequences (57) and that of entire human and bovine mitochondrial genome (2) has clearly demonstrated that the genetic code used in mitochondria is different from the "universal" code and, among other features, that the UGA termination codon specifies the amino acid tryptophan. Other differences seem to be peculiar to the mammalian mitochondrial genetic code, i.e., AGA and AGG are read as stop codons instead of isoleucine and the initiation codon can be either AUG or AUA or AUU. These results also demonstrate that there is a difference between mitochondria from different organisms due, perhaps, to the different pressures to which they have been subjected during evolution. These discoveries together with the finding that a small number of tRNAs is coded by mitochondrial genomes have allowed a better approach to the analysis of mitochondrial protein synthesis.

The first positive result of the correct *in vitro* translation of a mitochondrial mRNA in a heterologous system was reported by De Ronde et al. (21). They used a mitochondrial RNA fraction from *Saccharomyces carlbergensis* enriched in mRNA for the subunit II of cytochrome oxidase to direct protein synthesis in a wheat-germ cell-free system. According to the DNA sequence (28), five UGA codons that are dispersed throughout this mRNA have to be read as tryptophan to achieve the complete synthesis of the polypeptide. For this reason they supplemented the system with 40 μg/ml of suppressor $tRNA^{Ser}_{UGA}$ purified from another yeast species, *S. pombe* (46). The products of the synthesis were immunoprecipitated in consecutive steps with preimmune serum and anti-holo-cytocrome *c* oxidase antibody, using a goat anti-rabbit serum to recover the immune complexes. The electrophoretic analysis of the immunoprecipitates, when compared to the *in vivo* synthesized subunit II cytochrome oxidase and to the *in vitro* and *in vivo* products of two well-defined OXI-1 mutants, revealed that a full-length subunit II had been synthesized only in the presence of the suppressor $tRNA^{Ser}_{UGA}$; this confirmed that in mitochondria, UGA is not a termination codon as it is in prokaryotes, eukaryotes, and other organelle genomes.

Chloroplasts

As already mentioned in the introductory remarks, incorporation of radioactive amino acids into proteins of intact chloroplast has for several years been the only procedure followed by many laboratories to study protein synthesis in this system (11,63). The results obtained have led to the finding that in chloroplasts, as well as in mitochondria, some of the final products of synthesis arise from the assembly of subunits synthesized in separate loci: the cytoplasm and the chloroplast itself. The polypeptides known to be synthesized in the chloroplast are the large subunit of ribulose diphosphate carboxylase (17), the two largest subunits of the coupling factor CF1 (61), and a thylakoid membrane protein coded by chloroplast DNA from *Zea mays* (6). The use of *in vitro* protein synthesis systems with cloned pieces of chloroplast DNA will be of great help to locate on the chloroplast genome genes coding for polypeptides that, until now, have been genetically demonstrated to be of chloroplast origin. We will describe only few reports on the *in vitro* synthesis of chloroplast proteins using heterologous systems directed by chloroplast DNA or RNA, since they do not introduce any particular improvement to the experimental procedures already described for prokaryotes and mitochondria. Nevertheless, they demonstrate that the mechanism of translation of chloroplast mRNA is very similar to that of prokaryotes.

The work of Hartley et al. (38) demonstrated for the first time the existence in chloroplasts of an mRNA specific for the large subunit of the enzyme ribulose diphosphate carboxylase (RuBP case LS), which could be translated into a polypeptide of correct molecular weight using an *E. coli* cell-free system. They used total chloroplast RNA extracted from spinach plants to direct protein synthesis in an S-30 extract that could function correctly for the synthesis of the phage MS2 coat protein, as already reported by other authors. The major products synthesized after addition of chloroplast RNA were a 52,000 molecular weight polypeptide with several properties of the RuBP case LS extracted from isolated spinach chloroplasts and a smaller 35,000 molecular weight component probably analogous to a membrane-associated polypeptide that was also synthetized in isolated pea chloroplasts (24).

The synthesis of about 20 polypeptides is described by Bottomley et al. (12) using a coupled *in vitro* transcription-translation system as described by Zubay et al. (100,101). They adjusted their experimental conditions in order to obtain the lowest level of endogenous incorporation [^{35}S]methionine and the highest incorporation in response to added chloroplast DNA. This result was obtained by optimizing at 2 hr the preincubation period of the S-30 prior to the final dyalisis and at 14 m*M* the concentration of Mg^{2+}. Digestion of template DNA with restriction endonucleases demonstrated that the pattern of the products of DNA-directed protein synthesis could be altered. It also strongly indicated possible locations of cistrons in the chloroplast DNA. The complete DNA gene sequence of *Z. mays* RuBP case has been recently obtained (60) and compared to the sequence of some fragment of the purified peptide. The results of this analysis demonstrate, among other features, that the codon usage in chloroplast is similar to that found in human mitochondria, with the exception that UGA codon is a termination codon as in the "universal" code. The gene for this enzyme is also expressed in a polypeptide of correct molecular weight when cloned into *E. coli* (32).

EUKARYOTIC SYSTEM

When comparing the eukaryotic *in vitro* translation system to the prokaryotic one, we should keep in mind that the eukaryotic mRNA is more stable than the prokaryotic and, moreover, that eukaryotic mRNA contains a poly(A) tail attached to the 3′ end (thus being easily purified from ribosomal and transfer RNA by means of a poly(U) Sepharose or oligo(dT) cellulose columns). Finally, cloned DNA genes until recently were unavailable and DNA-directed translation experiments (which fail whenever introns are present) are therefore not reported.

The protein products are often identified by immunoprecipitation. Three *in vitro* translation systems are the most frequently used: the reticulocyte lysate (7,81,94), the wheat-germ (58,78,83) and the mouse ascites cell systems (3,59). The reticulocyte lysate is very active but contains a high concentration of endogenous globin mRNA. This can be removed (72) by treatment of the lysate with micrococcal nuclease (subsequently inactivated by addition of EGTA), leaving 70% of its original protein synthetic activity. This system contains very little, if any, nuclease or protease activity, and the activity of added mRNA is proportional to the measured amino acid incorporation. The wheat-germ system has very little endogenous mRNA, whereas the ascites cell extract can be preincubated to reduce the background activity. Both extracts can be freed of endogenous amino acids by Sephadex G-50 chromatography, thus increasing the specific radioactivity of the proteins synthesized. The reticulocyte system treated with nuclease shows a protein with a molecular weight of 42,000, which is labeled with [^{35}S]methionine even when an inhibitor of protein synthesis is used (71). The reticulocyte system treated with micrococcal nuclease appears to be the best presently available: in fact, mRNAs are translated up to 50 times with this system, whereas they appear to be translated only a few times with the others (13). Moreover, the reticulocyte system is more efficient in translating long mRNAs, especially if these are first denatured either by heat or by treatment with methyl mercury hydroxide (70).

These *in vitro* systems are active even when using a bulk RNA preparation in which mRNA may be as little as about 1% of the total RNA. Quite often the mRNA used is purified by means of poly(U) Sepharose [or oligo(dT) cellulose] columns, and sometimes further by size. The mRNA used should, of course, be undegraded, and a procedure involving homogenization in high concentrations of guanidine hydrochloride has proven to be very useful for isolating biologically active mRNAs of high molecular weight (20,86). The *in vitro* translation system should be optimized with respect to K^+ and Mg^{2+} ions, since each mRNA behaves differently (85,88).

The synthetic products of the eukaryotic *in vitro* translation system are often analyzed by SDS-polyacrylamide electrophoresis and autoradiography in order to establish the number of proteins synthesized and their molecular weight. The number of proteins synthesized may be used as a qualitative assay during purification of a specific mRNA. The identification of a translation product is often based on its reactivity with specific antibodies and, less frequently, on the peptide map and amino terminal amino acid sequence. Sometimes, the protein synthesized *in vitro* is a precursor of the finished protein: this conclusion is often based on a discrepancy between the expected and the observed molecular weight of a band reacting with a specific antibody. *In vitro* translation systems have been very important operationally in establishing the existence of precursor proteins. Proteins to be exported across the cell membrane often contain a short, hydrophobic sequence called the "signal" peptide (37). This peptide mediates the extrusion of secretory proteins through the endoplasmic reticulum membrane, and this process was studied by the use of an *in vitro* translation system in the presence of membrane preparations (22,45,82).

The *in vitro* translation system may also be used to identify the presence of mRNA coding sequences in a cloned DNA. One method, called hybrid arrested translation, is based on the inhibition of translation caused by DNA hybridizing to mRNA: if the cloned DNA is specific for the desired mRNA species, translation of that protein product will be inhibited by prehybridization with excess DNA (39,69). If the mRNA-DNA hybrid is heat-denatured prior to translation, the specific protein will reappear. The mRNA-DNA hybrid may be separated by chromatography from unhybridized mRNA (95), and in this case, only one translation product will be obtained after heat denaturation of the hybrid. Another way of separating the mRNA is by hybridization to either nitrocellulose filters containing bound cloned DNA (77) or diazobenzyloxymethyl paper containing covalently bound cloned DNA (84a,85a). Maintenance of protein synthesis in the reticulocyte lysate requires the addition of hemin (4). In the absence of hemin a translational repressor accumulates to block translation initiation. The action of repressor is prevented by 3′, 5′-cyclic AMP or 2-amino purine (52). Phosphorylated sugars have been reported to stimulate or inhibit protein synthesis in this system (52a,95a). In some cases, the reticulocyte system may be poorly efficient unless the proper tRNA is added. This may be related to the reported (84) deficiency of some tRNAs in reticulocytes. Encephalomyocarditis virus RNA is translated efficiently only if the *in vitro* system is supplemented with heterologous (rat liver) tRNA (72). Translation of silk fibroin mRNA will not take place unless tRNA from the posterior silk gland, which synthesizes fibroin, is added. Little or no stimulation of synthesis was obtained by addition of *E. coli* tRNA (56).

ACKNOWLEDGMENTS

We thank Dr. Richard Pau for revising the manuscript.

REFERENCES

1. Adams, J. M., and Capecchi, M. R. (1966): N-formylmethionyl-sRNA as the initiator of protein synthesis. *Proc. Natl. Acad. Sci. USA*, 55:147–155.
2. Anderson, S., Bankier, A. T., Barrell, B. G., de Bruijn, M. H. L., Coulson, A. R., Drouin, J., Eperon, I. C., Nierlich, D. P., Roe, B. A., Sanger, F., Schreier, P. H., Smith, A. J. H., Staden, R., and Young, I. G. (1981): Sequence and organization of the human mitochondrial genome. *Nature*, 290:457–465.
3. Aviv, H., Boime, I., and Leder, P. (1971): Protein synthesis directed by encephalomyocarditis virus RNA: properties of a transfer RNA-dependent system. *Proc. Natl. Acad. Sci. USA*, 68:2303–2307.
4. Balkow, K., Hunt, T., and Jackson, R. J. (1975): Control of protein synthesis in reticulocyte lysates: the effect of nucleotide triphosphates on formation of the translational repressor. *Biochem. Biophys. Res. Commun.*, 67:366–375.
5. Beattie, D. S., Lin, L.-F., and Stuchell, R. N. (1974): Studies on the control of mitochondrial protein synthesis in yeast. In: *The Biogenesis of Mitochondria*, edited by A. M. Kroon and C. Saccone, pp. 465–475. Academic Press, New York.
6. Bedbrook, J. R., Link, G., Coen, D. M., Bogorad, L., and Rich, A. (1978): Maize plastid gene expressed during photoregulated development. *Proc. Natl. Acad. Sci. USA*, 75:3060–3064.
7. Berns, J. M., and Bloemendal, H. (1974): Translation of mRNA from vertebrate eye lenses. *Methods Enzymol.*, 30:675–694.
8. Blossey, H. C., and Küntzel, H. (1972): In vitro translation of mitochondrial DNA from *Neurospora crassa*. *FEBS Lett.*, 24:335–338.
9. Bonitz, S. C., Berlani, R., Coruzzi, G., Li, M., Macino, G., Nobrega, F. G., Nobrega, M. P., Thalenfeld, B. E., and Tzagoloff, A. (1980): Codon recognition rules in yeast mitochondria. *Proc. Natl. Acad. Sci. USA*, 77:3167–3170.
10. Borst, P., and Grivell, L. A. (1978): The mitochondrial genome of yeast. *Cell*, 15:705–723.
11. Bottomley, W., Spencer, D., and Whitfeld, P. R. (1974): Protein synthesis in isolated spinach chloroplasts: comparison of light-driven and ATP-driven synthesis. *Arch. Biochem. Biophys.*, 164:106–117.
12. Bottomley, W., and Whitfeld, P. R. (1979): Cell-free transcription and translation of total spinach chloroplast DNA. *Eur. J. Biochem.*, 93:31–39.

13. Brawerman, G. (1974): Eukaryotic messenger RNA. *Annu. Rev. Biochem.*, 43:621–642.
14. Capecchi, M. R. (1967): Polypeptide chain termination *in vitro*: isolation of a release factor. *Proc. Natl. Acad. Sci. USA*, 58:1144–1151.
15. Chambers, D. A., and Zubay, G. (1969): The stimulatory effect of cyclic adenosine 3′,5′-monophosphate on DNA-directed synthesis of β-galactosidase in a cell-free system. *Proc. Natl. Acad. Sci. USA*, 63:118–122.
16. Chuang, D. -M., and Weissbach, H. (1973): Effect of eukaryote DNA on amino acid incorporation in extracts of *E. coli*. *Arch. Biochem. Biophys.*, 157:28–35.
17. Coen, D. M., Bedbrook, J. R., Bogorad, L., and Rich, A. (1977): Maize chloroplast DNA fragment encoding the large subunit of ribulosebisphosphate carboxylase. *Proc. Natl. Acad. Sci. USA*, 74:5487–5491.
18. Cooper, C. S., and Avers, C. J. (1974): Evidence of involvement of mitochondrial polysomes and messenger RNA in synthesis of organelle proteins. In: *The Biogenesis of Mitochondria*, edited by A. M. Kroon and C. Saccone, pp. 289–303. Academic Press, New York.
19. David, E. T., and Smith, K. E. (1981): Preparation and partial characterization of cell-free protein synthesizing extracts from *Tetrahymena pyriformis*. *Biochem. J.*, 194:761–770.
20. Deeley, R. G., Gordon, J. I., Burns, A. T. H., Mullinix, K. P., Binastein, M., and Goldberger, R. F. (1977): Primary activation of the vitellogenin gene in the rooster. *J. Biol. Chem.*, 252:8310–8319.
21. De Ronde, A., Van Loon, A. P. G. M., and Grivell, L. A. (1980): In vitro suppression of UGA codons in a mitochondrial mRNA. *Nature*, 287:361–363.
22. Dobberstein, B., and Blobel, G. (1977): Functional interaction of plant ribosomes with animal microsomal membranes. *Biochem. Biophys. Res. Commun.*, 74:1675–1682.
23. Dohan, F. C., Jr., Rubman, R. H., and Torriani, A. (1971): *In vitro* synthesis of *Escherichia coli* alkaline phosphatase monomers. *J. Mol. Biol.*, 58:469–479.
24. Eaglesham, A. R. J., and Ellis, R. J. (1974): Protein synthesis in chloroplasts. II. Light-driven synthesis of membrane proteins by isolated pea chloroplasts. *Biochim. Biophys. Acta*, 335:396–407.
25. Egberts, E., Traub, P., Herrlich, P., and Schweiger, M. (1972): Functional integrity of *Escherichia coli* 30-S ribosomes reconstituted from RNA and protein: *in vitro* synthesis of S-adenosylmethionine cleaving enzyme. *Biochim. Biophys. Acta*, 277:681–684.
26. Ehring, R., Beyreuther, K., Wright, J. K., and Overath, P. (1980): *In vitro* and *in vivo* products of *E. coli* lactose permease gene are identical. *Nature*, 283:537–540.
27. Firtel, R. A., Jacobson, A., and Lodish, H. F. (1972): Isolation and hybridization kinetics of messenger RNA from dictyostelium. *Nature [New Biol.]*, 239:225–228.
28. Fox, T. D. (1979): Five TGA "stop" codons occur within the translated sequence of the yeast mitochondrial gene for cytochrome *c* oxidase subunit. II. *Proc. Natl. Acad. Sci. USA*, 76:6534–6538.
29. Fuchs, E., and Fuchs, C. M. (1971): *In vitro* synthesis of T3 and T7 RNA polymerase at low magnesium concentration. *FEBS Lett.*, 19:159–162.
30. Fuchs, E. (1976): The interdependence of magnesium with spermidine and phosphoenolpyruvate in an enzyme-synthesizing system *in vitro*. *Eur. J. Biochem.*, 63:15–22.
31. Gamble, J. G., and McCluer, R. II (1970): *In vitro* studies with rifampicin on the stability of heart mitochondrial RNA. *J. Mol. Biol.*, 53:557–560.
32. Gatenby, A. A., Castleton, J. A., and Saul, M. W. (1981): Expression in *E. coli* of maize and wheat chloroplast genes for large subunit of ribulose bisphosphate carboxylase. *Nature*, 291:117–121.
33. Gold, L. M., and Schweiger, M. (1969): Synthesis of phage-specific α- and β-glucosyl transferases directed by T-even DNA *in vitro*. *Proc. Natl. Acad. Sci. USA*, 62:892–898.
34. Greenblatt, J., Li, J., Adhia, S., Friedman, I., Baron, L. S., Redfield, B., Kung, H. F., and Weissbach, H. (1980): L factor that is required for β-galactosidase synthesis is the *nusA* gene product involved in transcription termination. *Proc. Natl. Acad. Sci. USA*, 77:1991–1994.
35. Halbreich, A. (1979): Differential effect of Ca^{+2} on the translation of yeast mitochondrial and some viral RNA's in *E. coli* cell-free system. *Biochem. Biophys. Res. Commun.*, 86:78–87.
36. Haldar, D. (1971): Protein synthesis in isolated rat brain mitochondria. *Biochem. Biophys. Res. Commun.*, 42:899–904.
37. Hamlyn, P. (1977): Green light for the signal hypothesis? *Nature*, 267:207.
38. Hartley, M. R., Wheeler, A., and Ellis, R. J. (1975): Protein synthesis in chloroplasts. V. Translation of messenger RNA for the large subunit of fraction I protein in a Heterologous cell-free system. *J. Mol. Biol.*, 91:67–77.

39. Hastie, N. D., and Held, W. A. (1978): Analysis of mRNA populations by cDNA•mRNA hybrid-mediated inhibition of cell-free protein synthesis. *Proc. Natl. Acad. Sci. USA*, 75:1217–1221.
40. Hendler, F., Halbreich, A., Jakovcic, S., Patzer, J., Merten, S., and Rabinowitz, M. (1976): Characterization and translation of yeast mitochondrial RNA. In: *Genetics and Biogenesis of Chloroplasts and Mitochondria*, edited by T. Bucher et al., pp. 679–684. Elsevier/North-Holland Biomedical, Amsterdam.
41. Hoagland, M. B., Stephenson, M. L., Scott, J. F., Hecht, L. I., and Zamecnik, P. C. (1958): A soluble ribonucleic acid intermediate in protein synthesis. *J. Biol. Chem.*, 231:241–257.
42. Hsu, W. (1971): Translation of an RNA viral message *in vitro*: one step polypeptide chain elongation. *Biochem. Biophys. Res. Commun.*, 42:405–412.
43. Jacobs, K. A., and Schlessinger, D. (1977): *Escherichia coli* DNA-directed β-galactosidase synthesis in presence and absence of Ca^{2+}. *Biochemistry*, 16:914–920.
44. Jacobs, K. A., Shen, V., and Schlessinger, D. (1978): Coupling of *lac* mRNA transcription to translation in *Escherichia coli* cell extracts. *Proc. Natl. Acad. Sci. USA*, 75:158–161.
45. Katz, F. N., Rothman, J. E., Lingappa, V. R., Blobel, G., and Lodish, H. F. (1977): Membrane assembly *in vitro*: synthesis, glycosylation, and symmetric insertion of a transmembrane protein. *Proc. Natl. Acad. Sci. USA*, 74:3278–3282.
46. Kohli, J., Kwong, T., Altruda, F., and Söll, D. (1979): Characterization of a UGA-suppressing serine tRNA from *Schizosaccharomyces pombe* with the help of a new *in vitro* assay system for eukaryotic suppressor tRNAs. *J. Biol. Chem.*, 254:1546–1551.
47. Kozak, M., and Nathans, D. (1972): Translation of the genome of a ribonucleic acid bacteriophage. *Bacteriol. Rev.*, 36:109–134.
48. Kung, H. F., Eskin, B., Redfield, B., and Weissbach, H. (1979): DNA-directed *in vitro* synthesis of β-galactosidase: requirement for formylation of methionyl-$tRNA_f$. *Arch. Biochem. Biophys.*, 195:396–400.
49. Kung, H. F., Redfield, B., and Weissbach, H. (1979): DNA-directed *in vitro* synthesis of β-galactosidase. Purification and characterization of stimulatory factors in an ascites extract. *J. Biol. Chem.*, 254:8404–8408.
50. Lederman, M., and Zubay, G. (1967): DNA-directed peptide synthesis. I. A comparison of T2 and *Escherichia coli* DNA-directed peptide synthesis in two cell-free systems. *Biochem. Biophys. Acta*, 149:253–258.
51. Lederman, M., and Zubay, G. (1968): DNA-directed peptide synthesis. V. The cell-free synthesis of a polypeptide with β-galactosidase activity. *Biochem. Biophys. Res. Commun.*, 32:710–714.
52. Legon, S., Brayley, A., Hunt, T., and Jackson, R. J. (1974): The effect of cyclic AMP and related compounds on the control of protein synthesis in reticulocyte lysates. *Biochem. Biophys. Res. Commun.*, 56:745–752.
52a. Lenz, J. R., Chatterjee, G. E., Maroney, P. A., and Baglioni, C. (1978): Phosphorylated sugars stimulate protein synthesis and Met-$tRNA_f$ binding activity in extracts of mammalian cells. *Biochemistry*, 17:80–87.
53. Leventhal, J. M., and Chambliss, G. H. (1979): DNA-directed cell-free protein synthesizing system of *Bacillus subtilis*. *Biochem. Biophys. Acta*, 564:162–171.
54. Lindberg, U., and Persson, T. (1972): Isolation of mRNA from KB-cells by affinity chromatography on polyuridylic acid covalently linked to sepharose. *Eur. J. Biochem.*, 31:246–254.
55. Lizardi, P. M., and Luck, J. L. (1972): The intracellular site of synthesis of mitochondrial ribosomal proteins in *Neurospora crassa*. *J. Cell Biol.*, 54:56–74.
56. Lizardi, P. M., Mahdavi, V., Shields, D., and Candelas, G. (1979): Discontinuous translation of silk fibroin in a reticulocyte cell-free system and in intact silk gland cells. *Proc. Natl. Acad. Sci. USA*, 76:6211–6215.
57. Macino G., and Tzagoloff, A. (1979): Assembly of the mitochondrial membrane system: Partial sequence of a mitochondrial ATPase gene in *Saccharomyces cerevisiae*. *Proc. Natl. Acad. Sci. USA*, 76:131–135.
58. Marcus, A., Efron, D., and Weeks, D. P. (1974): The wheat embryo cell-free system. *Methods Enzymol.*, 30:749–754.
59. Mathews, M. B., and Korner, A. (1970): Mammalian cell-free protein sysnthesis directed by viral ribonucleic acid. *Eur. J. Biochem.*, 17:328–338.
60. McIntosh, L., Poulsen, C., and Bogorad, L. (1980): Chloroplast gene sequence for the large subunit of ribulose bisphosphatecarboxylase of maize. *Nature*, 288:556–560.
61. Mendiola-Morgenthaler, L. R., Morganthaler, J. J., and Price, C. A. (1976): Synthesis of coupling factor CF_1 protein by isolated spinach chloroplasts. *FEBS Lett.*, 62:96–100.

62. Miller, J. H. (1974): Cell-free synthesis of β-galactosidase. In: *Experiments in Molecular Genetics*, pp. 419–424. Cold Spring Harbor Laboratory, New York.
63. Morgenthaler, J. J., and Mendiola-Morgenthaler, L. (1976): Synthesis of soluble, thylakoid, and envelope membrane proteins by spinach chloroplasts purified from gradients. *Arch. Biochem. Biophys.*, 172:51–58.
64. Nathans, D., Notani, G., Schwartz, J. H., and Zinder, N. D. (1962): Biosynthesis of the coat protein of coliphage f2 by *E. coli* extracts. *Proc. Natl. Acad. Sci. USA*, 48:1424–1431.
65. Nierman, W. C., and Chamberlin, M. J. (1979): Studies of RNA chain initiation by *E. coli* RNA polymerase bound to T7 DNA. *J. Biol. Chem.*, 254:7921–7926.
66. Nirenberg, M. W. (1976): Cell-free protein synthesis directed by messenger RNA. *Methods Enzymol.*, 6:17–23.
67. Nirenberg, M. W., and Matthaei, J. H. (1961): The dependence of cell-free protein synthesis in *E. coli* upon naturally occurring or synthetic polyribonucleotides. *Proc. Natl. Acad. Sci. USA*, 47:1588–1602.
68. Padmanaban, G., Hendler, F., Patzer, J., Ryan, R., and Rabinowitz, M. (1975): Translation of RNA that contains polyadenylate from yeast mitochondria in an *Escherichia coli* ribosomal system. *Proc. Natl. Acad. Sci. USA*, 72:4293–4297.
69. Paterson, B. M., Roberts, B. E., and Kuff, E. L. (1977): Structural gene identification and mapping by DNA•mRNA hybrid-arrested cell-free translation. *Proc. Natl. Acad. Sci. USA*, 74:4370–4374.
70. Payvar, F., and Schimke, R. T. (1979): Methylmercury hydroxide enhancement of translation and transcription of ovalbumin and conalbumin mRNAs. *J. Biol. Chem.*, 254:7636–7642.
71. Pelham, H. R. B. (1978): Translation of encephalomyocarditis virus RNA *in vitro* yields an active proteolytic processing enzyme. *Eur. J. Biochem.*, 85:457–462.
72. Pelham, H. R. B., and Jachson, R. J. (1976): An efficient mRNA-dependent translation system from reticulocyte lysates. *Eur. J. Biochem.*, 67:247–256.
73. Perlman, S., Abelson, H. T., and Penman, S. (1973): Mitochondrial protein synthesis: RNA with the properties of eukaryotic messenger RNA. *Proc. Natl. Acad. Sci. USA*, 70:350–353.
74. Pouwels, P. H., and Van Rotterdam, J. (1972): *In vitro* synthesis of enzymes of the tryptophan operon of *Escherichia coli*. *Proc. Natl. Acad. Sci. USA*, 69:1786–1790.
75. Primakoff, P., and Artz, S. W. (1979): Positive control of *lac* operon expression *in vitro* by guanosine 5′-diphosphate 3′-diphosphate. *Proc. Natl. Acad. Sci. USA*, 76:1726–1730.
76. Rendi, R. (1959): The effect of chloramphenicol on the incorporation of labeled amino acids into proteins by isolated subcellular fractions from rat liver. *Exp. Cell Res.*, 18:187–189.
77. Ricciardi, R. P., Mille, J. S., and Roberts, B. E. (1979): Purification and mapping of specific mRNAs by hybridization-selection and cell-free translation. *Proc. Natl. Acad. Sci. USA*, 76:4927–4931.
78. Roberts, E. B., and Paterson, B. M. (1973): Efficient translation of tobacco mosaic virus RNA and rabbit globin 9S RNA in a cell-free system from commercial wheat germ. *Proc. Natl. Acad. Sci. USA*, 70:2330–2334.
79. Salser, W., Gesteland, R. F., and Bolle, A. (1967): *In vitro* synthesis of bacteriophage lysozyme. *Nature*, 215:588–591.
80. Schatz, G., and Mason, T. L. (1974): The biosynthesis of mitochondrial proteins. *Annu. Rev. Biochem.*, 43:51–87.
81. Schimke, R. T., Rhoads, R. E., and McKnight, G. S. (1974): Assay of ovalbumin mRNA in reticulocyte lysate. *Methods Enzymol.*, 30:694–701.
82. Shields, D., and Blobel, G. (1977): Cell-free synthesis of fish preproinsulin, and processing by heterologous mammalian microsomal membranes. *Proc. Natl. Acad. Sci. USA*, 74:2059–2063.
83. Shih, D. S., and Kaesberg, P. (1973): Translation of brome mosaic viral ribonucleic acid in a cell-free system derived from wheat embryo. *Proc. Natl. Acad. Sci. USA*, 70:1799–1803.
84. Smith, D. W. E. (1975): Reticulocyte transfer RNA and hemoglobin synthesis. *Science*, 190:529–535.
84a. Smith, D. F., Searle, P. F., and Williams, J. G. (1979): Characterisation of bacterial clones containing DNA sequences derived from xenopus laevis vitellogenin mRNA. *Nucleic Acids Res.*, 6:487–506.
85. Sonenshein, G. E., and Brawerman, G. (1977): Differential translation of rat albumin messenger RNA in a wheat germ cell-free system. *Biochemistry*, 16:5445–5448.
85a. Stark, G. R., and Williams, J. G. (1979): Qualitative analysis of specific labelled RNA's using DNA covalently linked to diazobenzyloxy-methyl-paper. *Nucleic Acids Res.*, 6:195–203.

86. Strohman, R. C., Moss, P. S., Micou-Eastwood, J., Spector, D., Przybyla, A., and Paterson, B. (1977): Messenger RNA for myosin polypeptides: isolation from single myogenic cell cultures. *Cell*, 10:265–273.
87. Sugiyama, T., and Nakada, D. (1970): Translational control of bacteriophage MS2 RNA cistrons by MS2 coat protein: affinity and specificity of the interaction of MS2 coat protein with MS2 RNA. *J. Mol. Biol.*, 48:349–355.
88. Tse, T. P. H., and Taylor, J. M. (1977): Translation of albumin messenger RNA in a cell-free protein-synthesizing system derived from wheat germ. *J. Biol. Chem.*, 252:1272–1278.
89. Watson, M. D., Wild, J., and Umbarger, H. E. (1979): Positive control of *ilvC* expression in *Escherichia coli* K-12; identification and mapping of regulatory gene *ilvY*. *J. Bacteriol.*, 139:1014–1020.
90. Webster, R. E., Engelhardt, D. L., and Zinder, N. D. (1966): *In vitro* protein synthesis: chain initiation. *Proc. Natl. Acad. Sci. USA*, 55:755–761.
91. Weteckam, W., Staack, K., and Ehring, R. (1972): Relief of polarity in DNA-dependent cell-free synthesis of enzymes of the galactose operon of *Escherichia coli*. *Mol. Gen. Genet.*, 116:258–276.
92. Wood, W. B., and Berg, P. (1962): The effect of enzymatically synthesized ribonucleic acid on amino acid incorporation by a soluble protein-ribosome system from *Escherichia coli*. *Proc. Natl. Acad. Sci. USA*, 48:94–104.
93. Wood, W. B., and Berg, P. (1964): Influence of DNA secondary structure on DNA-dependent polypeptide synthesis. *J. Mol. Biol.*, 9:452–471.
94. Woodward, W. R., and Herbert, E. (1974): Preparation and analysis of nascent chains on reticulocyte membrane-bound ribosomes. *Methods Enzymol.*, 30:746–748.
95. Woolford, J. L., Jr., and Rosbash, M. (1979): The use of R-looping for structural gene identification and mRNA purification. *Nucleic Acids Res.*, 6:2483–2497.
95a. Wu, J. M., Cheung, C. P., and Suhadolnik, R. J. (1978): Inhibition of protein synthesis by glucose 6-phosphate and fructose 1,6-diphosphate in lysed rabbit reticulocytes and the reversal of inhibition by NAD^+. *Biochem. Biophys. Res. Commun.*, 82:921–928.
96. Yang, H., and Zubay, G. (1973): Synthesis of the arabinose operon regulator protein in a cell-free system. *Mol. Gen. Genet.*, 122:131–136.
97. Yang, H., Zubay, G., Urm, E., Reiness, G., and Cashel, M. (1974): Effects of guanosine tetraphosphate, guanosine pentaphosphate, and β, γ-methylenyl-guanosine pentaphosphate on gene expression of *Escherichia coli in vitro*. *Proc. Natl. Acad. Sci. USA*, 71:63–67.
98. Zubay, G., Lederman, M., and De vries, J. K. (1967): DNA-directed peptide synthesis. III. Repression of β-galactosidase synthesis and inhibition of repressor by inducer in a cell-free system. *Proc. Natl. Acad. Sci. USA*, 58:1669–1675.
99. Zubay, G. (1973): In vitro synthesis of protein in microbial systems. *Annu. Rev. Genet.*, 7:267–287.
100. Zubay, G., and Chambers, D. A. (1969): A DNA-directed cell-free system for β-galactosidase synthesis; characterization of the *de novo* synthesized enzyme and some aspects of the regulation of synthesis. *Cold Spring Harbor Symp. Quant. Biol.*, 34:753–761.
101. Zubay, G., Chambers, D. A., and Cheong, L. C. (1970): Cell-free studies on the regulation of the *lac* operon. In: *The* lac *Operon*, edited by D. Zipser and J. Beckwith, pp. 375–391. Cold Spring Harbor Laboratory, New York.
102. Zubay, G., Morse, D. E., Jurgen Schrenk, W., and Miller, J. H. (1972): Detection and isolation of the repressor protein for the tryptophan operon of *Escherichia coli*. *Proc. Natl. Acad. Sci. USA*, 69:1100–1103.

Subject Index